Sigmund Ringeck's

Knightly Art of the Longsword

Sigmund Ringeck's
Knightly Art of the Longsword

David Lindholm
and Peter Svärd

Paladin Press • Boulder, Colorado

Sigmund Ringeck's Knightly Art of the Longsword
by David Lindholm and Peter Svärd

ISBN 1-58160-410-6
Printed in the United States of America

Published by Paladin Press, a division of
Paladin Enterprises, Inc.
Gunbarrel Tech Center
7077 Winchester Circle
Boulder, Colorado 80301 USA
+1.303.443.7250

Direct inquiries and/or orders to the above address.

Cover illustration and design by Peter Svärd

All photographs courtesy of Peter Johnsson

Visit our Web site at www.paladin-press.com

David Lindholm
Has an MA in Medieval archaeology and history from the University of Lund and works as a writer and archaeologist. He has been training in European swordsmanship since 1986 and has some experience in Western fencing, iaido, and kenjutsu. He is a member of ARMA (the Association for Renaissance Martial Arts) and directs the ARMA study group in Malmö, Sweden. David spends all his spare time with his telescopes, pondering the magnificence of the universe. He translated the text and wrote the interpretations and comments in the book.

Dedication:
Frida and my little cat, my two absolutes in life.
And for Sigmund Ringeck with gratitude.

Peter Svärd
Works full-time as an art director and illustrator for a Swedish Internet consultancy firm. Has been active in various medieval/ fantasy societies since the late 1980s, training with swords on and off for about 10 years. His spare time is spent with his family and on new projects such as making trollish rubber masks. He illustrated this book.

Dedication:
Anja and Tove, two princesses that light up my life.

Contributors

Peter Johnsson is a swordsmith born in 1965 in Uppsala, Sweden. Presently he does documentary work on historical swords in various European collections, and creates both exacting reconstructions and interpretations based on impressions and facts. Peter wrote about the sword edges and sharpening, and made the swords we used in this book.

Lena Strid is an archaeologist and osteologist who works in Sweden and Denmark. Always interested in history, the older the better, she has a special interest in the area of battlefield trauma and the interpretation of fighting from osteological material. Lena helped with the translation.

Contents

Introduction by John Clements 1
Translation and Transcription of Ringeck's Manuscript 7
Johannes Liechtenauer 9
Master Ringeck's Longsword 11
How to Use This Material 15

Ringeck's Book 17
The Sword 23
The Techniques 25
Vier Leger 27
Zornhaw 33
The Four Openings 49
Mutieren and Duplieren 51
Krumphaw 57
Zwerchaw 69
Feler and Verkerer 83
Schiller 91
Schaytler 99
Vier Versetzen 105
Nachrayßen 119
Absetzen and Durchwechseln 129
Ringen am Schwert and Durchlauffen 137
Das Schnitt 151
Zway Hengen 157
Sprechfenster 161
Lang Ort 163
Beschließung der Zedel 169
Nebenhut 173
Absetzen 189
Schrankhut 191
Das Redel 197
Bruch 199
The Modern Practice of Swordsmanship 205
Physical Training and Swordsmanship 207
The Interpretation of Historical Sources 209
What is an Edge? 213
How to Sharpen a Sword 221
Test-Cutting 225
The Use of Armor 227
Glossary 229
Bibliography 235
Online Resources 237

Acknowledgments

This book is the result of the efforts and goodwill of many people, and we would like to express our gratitude to them. John Clements, who put us in touch with the publisher, helped with editing, and was an inspiration as well as one of the first serious modern exponents on European swordsmanship. Jörg Bellinghausen, for valuable insights and for sharing his translation of Sigmund Ringeck into English. Christoph Kaindel, for sharing his translation into modern German, and very valuable comments. Martin Wierschin, for his exemplary work on Liechtenauer, which made it publicly known. The work and efforts of these people have been very important for us and their work indicated some solutions in working with this material. We are indebted to them.

We would also like to thank Andreas Ingefjord, Mattias Ånmo, and Jonas Nilsson for dying well in the illustrations. Peter Buskas, for correcting our English and also for dying well. Lena Svärd, for correcting all those invisible errors. Teachers and professors at the departments of archaeology and history at the University of Lund for their encouragement and patience. Also, all the members of the growing group of people who practice and take an interest in the martial legacy of Europe. We all help each other to recover a lost history. Lastly, great thanks to Lena Strid who helped on the initial translations and has been an invaluable help in the work.

By a twist of fate, when we were halfway through our work we found out that we were not alone in our endeavor. It turned out that Mr. Christian Tobler had been working on a translation as well. We would therefore here like to compliment Mr. Tobler on his excellent work and recommend it to you good readers. Study these books and come up with all the things that we have missed. We hope that you all will find something of value between these covers.

David Lindholm and Peter Svärd
Malmö, Sweden, August 2002

Introduction by John Clements

He that knows how to fence will now understand these teachings.
–Master Sigmund Ringeck, 15th century

MASTER RINGECK SPEAKS AGAIN

THE STUDY OF historical European martial arts is undergoing an unprecedented revival, and earnest interest in actual medieval and Renaissance fighting skills has exploded in recent years. Much of what we know of fighting treatises from the Middle Ages is expanding and changing every year, and the old, "orthodox" view of fencing history is fading in the face of recent advances in scholarship and practice. The popular myth of untutored knights clumsily swinging crude swords while lumbering around in heavy armor is shredded by the actual evidence. The unequivocal picture presented by historical sources is one of trained warriors expertly employing skillfully designed weapons with brutal efficiency.

ALTHOUGH STILL IN its infancy, serious modern study and interpretation of the teachings of European martial arts literature is now well under way. While the term "martial arts" today is typically synonymous with "Asian fighting arts," for centuries highly sophisticated European martial arts existed. It is from the Latin that we actually derive the English term, "martial arts"–from "arts of Mars," the Roman god of war. Books and manuscripts on personal combat skills by European authors flourished in the 15th and 16th centuries. These little-known treatises and guidebooks on fighting skills are numerous and diverse. One of the most important today is that of *fechtmeister* (fencing master) Sigmund Ringeck.

RINGECK'S TEACHINGS ON the longsword are based on the *fechtbuch* (fencing book) of the great Swabian master, Johannes Liechtenauer. Liechtenauer's writings from the 1380s reflected a fencing method that had been developed by at least 1350. Liechtenauer was highly influential among other German fencing masters for the next two centuries, and his teachings were expanded and written upon by many throughout the 1400s and early 1500s. The techniques are typically described as examples of general actions to employ as ideal responses to various attacks. But, as Ringeck tells us, Liechtenauer "wrote down his teachings in secret and hidden words so that the art would not become known by all." Additionally, due to the character of their craft, medieval masters of defense often wrote in verse form when transmitting their lessons. When combined with the subtle nature of fencing instruction and the passage of time, the result produces somewhat cryptic passages that must be scrutinized at length. Because of this we must rely on later masters who interpreted his words.

THE BEAUTY OF Ringeck's commentaries on Liechtenauer is that his method is fairly simple. He describes core principles from which fundamental concepts are developed. A number of basic techniques follow from these and can, in turn, be employed in numerous variations. For example he tells us, "You shall always be ready and trained in the art of fencing so that you can use all counters against your opponent's techniques. From each counter you shall work with the three wonders." In the German school of medieval fencing these "three wonders," or *drey wunder*, were the attacks of the cut, the thrust, and the slice.

The definition of fencing in Ringeck's age was the art of using all weapons, but particularly swords, on foot or mounted, armored or unarmored, and always including grappling and wrestling techniques. We do not know much of the man Ringeck himself except that he was a knight and, at the time he produced his treatise, the fencing master to the Duke of Bavaria. As such, he himself would surely have seen combat as well as his fair share of tournaments and possibly judicial duels (as observer, if not participant). Paraphrasing Liechtenauer, Ringeck reveals the philosophical foundation of his teachings when he advises his readers, "Young knights learn to love God and honor women. Be chivalrous and learn the art that your honor will increase in war. Wrestle well, skillfully wield spear, sword, and dagger in a manful way."

While Ringeck's material on the longsword is not comprehensive, it is considerable and provides a view of core principles and concepts crucial to understanding how the weapon was really used. His teachings consist essentially of fundamental countertechniques to fundamental attacks that can then be used in endless variations. The approach of this modern version uses artwork to illustrate and interpret his teachings. The advantage of this approach over photographs is to make each example less subject to personal idiosyncrasies while being able to convey an appropriate sense of motion and inertia in each action. This is the most accurate transcription and translation of Ringeck's *fechtbuch* yet available. While by no means a definitive interpretation, as this subject is open to continual refinement and reexamination, it will be a boon for the student of medieval fencing. A translation of the material alone is not as useful as a reliable interpretation based on hands-on martial study. This book is just such a work. For too long this material has remained hidden or obscure, and too frequently modern enthusiasts of this subject have had only limited experience with the historical source literature. They have also lacked the physical experience and scientific skills by which to undertake its research. This situation is finally changing.

But in the interpretation of forgotten combatives there can be the danger today of misconceptions formed as a result of assumptions made without specialized knowledge or understanding. This process is invariably colored by our own ideas and beliefs about the functioning of historical arms. This modern interpretation of Ringeck, however, is a well-rounded study derived from extensive experience, not just in mock-combat reenactment or martial sport play, but the benefits of test-cutting with sharp swords and examination of antique sword specimens.

It is not difficult to note that a good deal of the medieval (and Renaissance) combat reconstruction that has taken place in recent times has been based on assumptions that are decades old. There has not been very much reliable reference material, and efforts have, therefore, tended to borrow from sport or theatrical fencing or from Asian arts, with a good portion simply being made up. Modern students will therefore greatly benefit from this material, sparking experiment, practice, new insights, and reconsideration. Having this material to study prevents our having to work blind, helps us avoid the dead-ends and trial and error approach and lets us reexamine our own ideas about the wielding of European longswords. As we do this, we must be aware that whenever historical fighting methods–devised to actually kill and maim with real weapons–are applied for purposes of pretend-performance or scoring points in sporting play, there is a significant and profound change that must occur as a result. There is a definite difference between interpreting and reconstructing medieval fencing techniques for the objective of practicing them as deadly skills, as opposed to performing them as choreographed displays, or in sporting competitions. When students examine the instructions of the old masters in light of what can be realistically faked for an audience or what is allowed "under the rules," it changes the entire outlook of how and why they practice in the first place. It also affects understanding and utilization of the core principles. In the modern reconstruction of historical European martial arts, the significance of modifying these teachings as you interpret cannot be overstated.

The approach taken within this book is to explain the original meaning of Ringeck's teachings for the purpose of safe practice. Fortunately for modern practitioners of medieval fencing, Ringeck divulges the flexibility of using his method when he writes, "lords and gentlemen learn how to survive in earnest and in play by this art," indicating that it was suited both for actual combat and, with some modifications, for practice fights.

Yet, any modern reconstruction or interpretation will inevitably be incomplete because we cannot know from historical materials alone the exact totality of any forgotten combat art. The whole problem in studying or practicing any martial art style is how you determine if something would really be combat effective. With little or no understanding or experience in actual hand-to-hand combat using medieval weapons today, this issue is often hotly debated. Questions of what would or would not readily work in real fighting are often reduced to subjective impressions. As modern students, our objective must be to reduce this element and rely on more pragmatic and empirical views of what the historical sources meant in their instructions at the time. We must endeavor to advance our understanding and continually question our assumptions. Our goal should always be to avoid limited, predictable, or stylistic technique in our understanding, and instead to be open, inclusive, and constantly refining our skill.

Historical fighting manuals such as Ringeck's are our first and most authoritative source for information on how medieval warriors employed their weapons. When it comes to discovering just how these swordsmen learned and practiced their skills, though, manuals such as Ringeck's leave many questions unanswered. They do not tell us, for example, at what speed they practiced attacks and counterattacks, or at what level of force or degree of range was commonly used when doing so, nor do they specify the necessary grip changes for different actions, or even what portion of your blade contacts what portion of the opponent's blade. They do not describe how they chose a good blade or sharpened its edges. These things have to be rediscovered on our own. The manuals are the primary source, but they must still be interpreted through other research.

The techniques developed and taught by the masters of defence like Sigmund Ringeck were no mere theories. They were not collections of "tricks" based on brute strength or the heaviest armor. They instead were moves that fighting men knew worked in combat, that they had discerned, had named, and had systematically taught to others. They actually applied them. In the brutality of medieval combat the effective was quickly separated from the ineffective, and only functionally practical elements were retained. But today, we don't "apply" them because we don't engage in lethal combat with sharp weapons. So, when we look at their instructions of how to fight or defend ourselves with certain weapons, we are obviously doing so from a very different perspective. Arguably, we do not yet fully understand the teachings of the medieval masters even as we try to earnestly employ them in mock-fighting.

Very often, when it comes to recent conceptualizations of medieval fencing, a number of different sources have all contributed misrepresentations of what it was all about (i.e., stage combat theories, theatrical performances, stylized rules of historical societies and reenactment groups, as well as fantasy role-playing games–both live and computerized–and the prejudices of modern sport fencing). These prevailing views stand in contrast to much of the teachings of the real historical masters and to much of the content of this book. In reading this work it will become unmistakably clear this is not the familiar swordplay of sport saber fencing, or the swordplay of movies and television, or of Japanese kendo or kenjutsu. The teachings of medieval longsword fencing may have similarities to other forms of fencing, but it is also its own distinct craft with many of its own unique elements.

This modern interpretation of Ringeck acquires its value from an awareness of the handling and function of actual historical longswords and an appreciation for the lethality, brutality, and elegance of the practice. The medieval longsword as employed by Ringeck and his fellow masters of defence was a versatile fighting style. It was brutal, but practical and efficient. It was a weapon of war as well as judicial duel, tournament, and, when necessary, personal self-defense. By the 14th century it was becoming the foundational weapon for fencing study. Such weapons, weighing from under 2.5 to just over 3.5 pounds, were agile, well-balanced, and superbly made cutting tools. They were not at all the ponderous clumsy things too often portrayed in contemporary popular culture. They were not wielded in the "hack and slash" style so familiar in movies, television, and video games. They were used in a tighter, closer manner that emphasized thrusts, employed both the forward and back edge in cutting, and utilized the hilt to trap or bind. Using both hands on longer blades allowed for stronger blows against more heavily armored opponents. When swung with both hands, the swords were capable of delivering devastating wounds.

These elements are reflected in Ringeck's teachings. On examination we find frequent use of winding and binding the sword against the opponent's blade, repeated use of counter blows to parry, and numerous examples of grappling, striking with hilt or pommel, use of "half-sword" gripping positions, plus arm grabs and blade seizures. Such actions, found throughout medieval fighting manuals, may seem unfamiliar when compared to Hollywood sword fights and other popular (mis)conceptions of medieval fencing, but they are the very essence of longsword fencing.

There are numerous fighting postures (or ready positions) used with European longswords, and different masters with different styles preferred some to others. But there are four basic, almost universal, stances among the many German and Italian masters and Ringeck presents these. These four fighting postures refer more or less to "ready positions" of holding the weapon to the outside, in the middle, aimed up high, or pointed down forward. Each of the basic wards, or guards, offers its own advantage of offense or defense. A fifth stance often found is with the weapon aimed low toward the back. There are also a great variety of secondary stances, as well as specialized "half-sword" positions found throughout the martial arts literature of the period. These adaptable guards, consisting of various ways of gripping the sword by the blade itself in either one hand or the other, underscore the diversity of the weapon. They take further advantage of the longsword's design to allow a fighter to deflect or stifle attacks and close in to stab with the point, beat with the hilt, or rake an opponent with the edge. Various half-swording positions were suited for fighting at different ranges and would be applied according to the temperament and physical disposition of the individual swordsman. They were particularly useful in fighting against heavy armor; however, they were employed whether the swordsman was in armor or not. These advanced techniques make up of some of the more challenging aspects of learning to fight with the weapon.

Ringeck also explains the use of the master cuts–vertical, diagonal, and horizontal blows that simultaneously defend and attack. This produced a dynamic and formidable method of combat that took full advantage of the longsword's versatility. What cannot be found in Ringeck's material, however, is the familiar "edge-on-edge" parrying so ubiquitous in modern depictions of swordplay and taught in the 18th and 19th centuries. Instead, instructions for defense stress protecting through your stance or ward and the "displacement" of attacks–deflecting an opponent's blow by the use of counter-strikes or avoiding the blow altogether–rather than blocking it directly with the sword. Thus "active" displacements deflect with the edge against the opponent's flat, while "passive" displacements receive his edge on your flat. Master Ringeck writes repeatedly of "breaking strikes" with other strikes. In fact, it is these very actions of counterstriking against the adversary's attacks that make up the large part of his entire method of fighting. Ringeck expresses this repeatedly. He says, for example, to, "strike him with the *krumphaw* against his sword," indicating a deflecting countercut against the flat of the their blade. This is not to be confused with his use of the edge of the blade to bind or to close, and greet or stifle the opponent's blow before it's fully committed.

When he writes, "You shall learn to displace strikes and thrusts with skill, so that his point is broken and yours hits" he is in fact referring to single-time actions that simultaneously "parry and riposte," the opposite of the slower "double-time" action of doing each separately. He further tells us, "There are four displacements which are very hurtful for the four guards, beware of displacements they [will] trouble you greatly." By this he means to be aware that the opponent will intercept your attacks with his own counterblows. What becomes clear, as we study more closely the manuals of such medieval fighting masters as Ringeck, is that these fighting arts did not separate the idea of attacks from defense nor separate fighting stances from either strikes or parries. In stating, "This is the first tenet of the longsword: learn to strike properly from both sides so that you learn to fence well and with strength," Ringeck teaches us that delivering proper blows from either the right or left (high or low) is the essence of both offense and defense. Ringeck was clearly not concerned with being "defensive." He instructs, "Do not concern yourself with what your opponent is doing: fence securely and you will strike so well that he will not get through with his own techniques." He wisely warns us against trying only to counterfight, saying, "All fencers that rely on displacing will be deceived and defeated with the feint." When Ringeck later adds, "Beware of all displacements used by bad fencers," he is not warning against using displacements themselves, but against imitating those wide, sweeping parries used by poor fighters (presumably, edge on edge).

While the German school of fencing in the Middle Ages taught that all attacks occur in one of three "times"–that is, before, during, or after the opponent's own action, Ringeck stated that it is the "middle" time (*indes*) that is most important. He said, "*Indes* is the key with which the art of fencing is unlocked." This can be understood in the way actions such as beating or binding on the opponent's blade, closing in, initiating disarms or wrestling, and resorting to half-sword techniques, are all executed not "before" or "after" the opponent's actions but "simultaneously" with them. Whereas striking before preempts the adversary's action and striking after exploits his action, "middle-time" strikes are timed to interrupt and intercept his actions.

The Langenschwert (longsword) as used by Ringeck and his fellow swordsmen was queen of weapons in its time. Although no traditional schools of medieval fencing survive to this day, modern enthusiasts are hard at work reconstructing and reviving the fighting skills represented in the old manuals. Yet, the activity of studying historical European martial arts today will by its nature always remain largely theoretical. This is obviously because no one can study genuine historical systems of defense in their original incarnation–no one today fights to the death with archaic weapons, nor is there anyone alive who was taught by someone who had. The intervening centuries have left us with only relics of the art. Because of this, study of the medieval longsword must consist of earnest physical training with accurate replica weapons combined with serious scholarship of the surviving martial arts literature.

Given the scarcity of material on this subject for most of the last two centuries, the longsword material in Sigmund Ringeck's *fechtbuch* is a welcome resource. It offers an alternative translation and interpretation reflecting the richness and diversity of not only the source material itself, but also the current community of historical fencing students. Both experienced practitioners and newcomers alike will benefit from this extensive analysis of the teachings of a 15th century master at arms. From this the reader will not find it difficult to begin his own extrapolations. Perhaps the most important piece of advice to consider when studying this work then, is to use it to reexamine your preconceptions, not reinforce them. This is an important work for historical European martial arts studies. As a researcher, practitioner, and teacher of medieval and Renaissance fencing, I am proud to have contributed to as well as benefited from it. I have no doubt it will prove a significant and exciting resource.

Having studied the craft since 1980, John Clements is one of America's leading instructors and practitioner-researchers of medieval and Renaissance fencing. He has researched swords in eight countries and taught seminars in six. He is director of ARMA (the Association for Renaissance Martial Arts) and author of the groundbreaking books Renaissance Swordsmanship *and* Medieval Swordsmanship, *both from Paladin Press.*

The Translation of Ringeck's Manuscript

All books have a history, and this one is no exception. It began with frustration at the very short supply of translations and illustrated interpretations of medieval and Renaissance fighting manuals. In short, it was a desire to try to do something that would benefit the enthusiasts of the sword, with the additional hope that others would find it a good idea and embark on similar projects.

We first came in contact with Ringeck's *Kunst des Fechtens* through the Internet, where portions of the manuscript from a translation by Jörg Bellinghausen were posted in English on the ARMA Web site in 1999. Previously our longsword practice had been based primarily on medieval and Renaissance illustrations in books and on murals, but with no real basis in the method of the use of the European longsword as perceived by the late medieval masters. We started to work on the material in 1999, but restricted ourselves to the material regarding the longsword for the moment. After working with the manuscript for some time we decided to acquire a copy of the original Ringeck text and also bought a copy of Döbringer from the Germanisches Nationalmuseum in Nürnberg. We soon found out that others were doing the same and we invited Mr. Bellinghausen to Sweden for a seminar in the spring of 2001. Thanks to his clarification of some important points we could complete the translation of the original manuscript by Ringeck, taking it even further.

The work has been done in two steps, the first being to make a complete transcription of the longsword material. This proved a daunting task at first, but we discovered that medieval German, curiously enough, is not unlike Swedish. And although David Lindholm had studied German for only six years in school, we found the text rather accessible (much more than modern German, we might add). Naturally there were difficult parts, and Martin Wierschin's dissertation (Wierschin 1965) on the Ringeck text proved an indispensable source of reference for obscure points both in the transcription and translation. At that time we received a copy of Christoph Kaindel's translation of the text into modern German, which provided another source for comparison on obscure points. To compare the different interpretations of other scholars is essential in this kind of work, since they offer variations in analysis and interpretations.

On completing the translation and transcription we picked out some key concepts and made comparisons with later manuals by Codex Wallerstein (c. 1450), and those by Hans Talhoffer (c. 1440s), Joachim Meyer (1570), and Jacob Sutor (1612). Primarily we did this to get a perspective on how some of the other masters chose to describe difficult concepts in illustrations and text. In short, this is a work that has come about through the efforts of others, both in the past and present.

In creating the modern English version of the text, we chose not to translate the original German text word for word, as we considered that beyond our capacity. However, we decided to keep the "merk verse" in a rougher style that better gave the feeling of the 14th century when they were compared. We have compiled the German text into a usable English version, making sure that all the technical terms are included. We have excluded some of the nontechnical text in the translation, but the whole original text is included for reference. Thus, the English version should be considered an interpretation within the limits of our present understanding.

We considered using photographs for illustrations, but decided against them to avoid personal idiosyncrasies. Instead we took photographs and converted each into a drawing.

Johannes Liechtenauer

Very little is known about Johannes Liechtenauer. He may have been born in Lichtenau in Franconia, Germany, around 1320 – 1330. He seems personally to have spawned the entire German fencing tradition of the longsword as it has come down to us today, but there were others involved through the course of history such as Ott, Lecküchner, Talhoffer, Meyer, Paulus Kal, and Peter von Danzig, each contributing something to the complete art as it was passed on down the centuries. But Liechtenauer stands out as the unrivaled originator of a system of combat focusing on the longsword, but with techniques on spear, sword and buckler, and unarmed combat. (We will not expound here on the interconnections between these masters. Those wishing a readable in-depth analysis can turn to Hans-Peter Hils, 1985.) Liechtenauer's ideas were recorded, or rather taught, in rhymed verse, a very common way in the Middle Ages when you were working with pupils that had no skills in writing or reading and who were unable to put your words on paper. It was also used by teachers who did not possess reading and writing skills, and we must remember that there is no direct evidence for assuming that Liechtenauer invented this system himself. Most likely he was just the only propagator that history has recorded, and the verses may have been his own illiterate way of memorizing. Be that as it may, his effort and influence remain undeniable.

We have no contemporary records of Liechtenauer, and the first historical reference is provided by the priest Hanko Döbringer in 1389. His text is the first to expound on the system of Liechtenauer. Ringeck's work is compiled about 50 years later and dated to the first half of the 15th century. That is the work we are concerned with here: the book of fencing by Sigmund Ringeck. Regarding Ringeck there has been an argument put forth that he could have been of the nobility, but this has not generally been given much credibility (Hils 1985). What we do know is that Ringeck, for a time at least, was employed as the fencing master to Albrecht, Duke of Bavaria. It is possible though that Ringeck may have been a member of the unfree lower knightly class that frequently acted as officials and teachers to the higher nobility in Germany (Barber 2000). Lacking more supporting evidence, there is little more to say on the matter. But it is a hypothesis that neither Liechtenauer nor Ringeck was alone or unique, and that there existed a broader system of teaching that had the opportunity to reach a larger part of the population. The Germans were well known during the Middle Ages for using larger swords than those used elsewhere in Europe. An interesting venue to investigate would be the Ordens Staat, the state created by the order of Teutonic Knights in the Baltic region, whose champions at least employed larger swords in duels and combat. And because many knights in Germany signed up for shorter spells of service in the 13th and 14th centuries, it may have provided a common ground. Another explanation may be found in the tournaments, an occupation and pastime that was very popular and to which nobles traveled long distances to fight. This would also provide a suitable environment for developing what Ringeck terms the knightly art of the longsword as opposed to other, more common weapons.

Note: There is a picture of Liechtenauer in Peter von Danzig's book. If that is to be taken as any form of evidence of how he looked, we come to an interesting conclusion: He was from Eastern Europe. There is no evidence that the kind of beard or fur-trimmed clothing of that cut were common in Western Europe. But in Eastern Europe, Poland, Bohemia, Russia, and the Baltic states, such appearance was normal.

Master Ringeck's Longsword

THE IMPORTANCE OF Sigmund Ringeck's book can hardly be overstated. It provides a link between the Middle Ages and the Renaissance in the art of longsword fighting. It is also one of the few surviving clear examples of the core teachings that constitute the foundation for the later German schools of the longsword (for an interesting suggestion on the relations between masters and manuscripts, see Hils). As with all the later German longsword fighting, the style is simple, although later masters made additions that gradually complicated the system. The simplicity of Ringeck's (or Liechtenauer's) method argues strongly in favor of its being a combative system rather than one of the later fencing school systems that existed within a less martial context.

RINGECK'S TEXT DOES raise several points of discussion, mainly due to the brevity of his writings at some crucial points, and the fact that he presupposes the reader having a good knowledge of the art of fencing beforehand. Therefore there is some debate today on how to interpret some of the more obscure points. A source of confusion is that we compare many different kinds of material and manuals to form a picture without having the scientific skills to evaluate the material in question–skills such as knowledge of the varying degrees of reliability of medieval and early Renaissance illustrations.

IT IS CLEAR that Ringeck, to some extent at least, is dependent on Döbringer's earlier text, which constitutes the earliest reference to Liechtenauer's system of fencing. But we are not in a position to evaluate to what extent Döbringer may have edited the text. Neither do we know if he made any inclusions or exclusions. Arguably his text would take precedence over Ringeck's due to its early appearance, and later works digressing from his should perhaps be attributed to a later version published sometime between Döbringer and his successors.

THERE ARE SOME topics in the text that arise from technical complications in interpreting Ringeck, and we would like to address some of these issues here rather than later in the text.

EDGE OR FLAT

RINGECK MAKES USE of both the edge and the flat in deflection, which makes good sense. What he does not do is tell you to make straight 90-degree parries edge on edge. As a matter of fact, if done correctly, a proper counterstrike will hit edge on flat, or even flat on edge, and place your edge at roughly 45 degrees to your opponent's sword. Although Ringeck does not specifically say to deflect by hitting against either edge or flat, striking edge on flat or flat on edge both work very well. If done with force, the blades might get small nicks, but if the fighting is in earnest this is of no consequence.

RINGECK STATES THAT you should strike against your opponent's blows to deflect them, and that the four *versetzen* are four strikes. The only instance where Ringeck actually tells you (not in words but by necessity of logic) to use the flat is the *absetzen*, and here you are told to wind your blade against your opponent's in a turning motion, not a flat push to the side. Otherwise it is the edge that is applied. There are several longswords in collections that can be accepted as having come from battlefields because they all have rather severe nicks in their edges (see photo).

AS A LAST point in this particular debate, the following might be worth considering. If we fight in armor, our swords will be nicked and dulled (historical sources state that this was a common occurrence) by hitting an opponent's armored body–if not his armor then by his weapon or the weapons of other opponents in a battle. Swords, like other weapons, were literally used up, which is also a simple explanation of why we have so few that have been used on the battlefield.

This sword blade shows some nicks, but admittedly we do not know how they got there.

FOOTWORK

RINGECK IS RATHER quiet on the subject of footwork, the reason being either that he expects his reader to be familiar with it already, or that he does not use it in the way we expect. There is a caveat on footwork–we are accustomed to the footwork of Asian martial arts, which are often unarmed, and of modern fencing, for example, but the mechanics of those fighting methods are different than those used when fighting with a longsword. Ringeck does not tell us only to void an attack (i.e., remove the body from an attack), but to counterstrike and void. It indicates that it is actually a lot faster to strike at an attack and then counter from the bind (contact) than it is to void it and then launch an attack. (We may here refer to George Silver and his true and false times.)[1] It is faster to move the hands than the body, which explains why Ringeck places such importance on both *duplieren* and *mutieren*, neither of which can be done without the bind. It is, however, important to remember that Ringeck gave very little instruction on how to fence with strikes, that is, at the middle range rather than the distant or close ranges. Ringeck deals almost exclusively with what you do when either party closes in into the next range. Perhaps this is because the middle-range fight is something that is rather simple and would include voiding. A closer distance makes voiding difficult, but Ringeck does nevertheless include it in some techniques such as *durchlauffen*. Ringeck, when mentioning footwork, does indicate that you should step away from your opponent's strike and counter in one movement, or step into the attack, stifle it, and then counter. While cautioning against too much movement, it is important never to remain still in fencing; footwork is essential.

ARMORED OR UNARMORED COMBAT

THE TECHNIQUES OF Ringeck's longsword are probably based on the use of the weapon in armor rather than without. This assumption is based on the fact that in the high Middle Ages the longsword could be classified primarily as a knightly weapon, at least up to the 14th century, and therefore its original use would probably have been primarily by someone in armor. (To counter my own statement, I would like to point out that there exist in Sweden pictures from the mid-14th century of men fighting with the longsword but wearing no armor.) So from an early date there definitely existed a tradition of using this weapon without armor.

IT IS ALSO interesting that the longsword, or the sword in general, was not the primary tournament weapon; this role belonged more to the poleax, hand ax, and lance and spear. Of course the sword was used at tournaments, but not as the primary weapon. This seems to make the longsword more of a weapon of war, self-defense, and judicial combat. Later it became endowed with great symbolic social significance. The nobility of the high and late Middle Ages did not indulge in unarmored combat very often: there was neither reason to fight nor the opportunity to do so. Therefore it is a reasonable assumption that the origin is to be found in armored combat, whether in war or perhaps the tournament.

WE MUST THEREFORE ignore Ringeck's seeming to make a distinction between fighting with or without armor. Rather, his armored combat has a very special context: a regulated judicial duel using predetermined weapons–longsword, spear, and dagger. But the longsword techniques described in the main section also work very well in battle, and the direct strikes against attacks make better sense since one cannot move as fast in armor. There is no time to void as often, and counters tend to be made in a more or less straight line.

IN THIS CONTEXT we should think about when you would use Ringeck's fighting art–not only on the battlefield, but in tourneys, duels, and in street quarrels. In pitched large-scale battles, of which there are many descriptions, there was probably neither space, time, nor sanity enough to indulge in the art of fencing. It is the pole arm that totally dominates the battlefield, along with the bow and, to a lesser extent, the crossbow. Perhaps we can see the later 15th century popularity of the longsword in fencing schools as an attempt by the city-dwelling man to acquire mastery of a symbol that earlier was primarily in the domain of the nobility. (This phenomenon is well-documented in the fields of clothing and social customs of the developing bourgeoisie, a form of social advancement in which lower classes gradually absorb symbols and values that used to belong to the upper class, no matter how we choose to define those groups.)

NOTE: IT IS important to distinguish between doing slow techniques and attacking at full speed and force, both in and out of armor. Both you and your weapon move very differently.

[1] George Silver's (*Paradoxes of Defence*) concept of true and false times is explained in detail in Terry Brown's *English Martial Arts*, pp. 93–97.

Master Sigmund Ringeck

No known portrait of Master Ringeck exists, so we made our own image of what he might have looked like.

How to Use This Material

WE INTENDED TO make this a book for use in modern study rather than reading in the comfort of your living room. It is meant to be a manual of suggestions, giving you ideas on how to work with Ringeck's style of longsword fighting. But it is not meant as the final word in any sense; that would be an unscientific delusion of grandeur on our part. Today we are all still very far from actually being able to lay claim to any degree of certainty in the understanding of the European fighting arts. But we have come a very long way in the past decade, mainly due to the efforts of a handful of people scattered over the globe.

THE TEXT

WE DECIDED TO follow the exact order of techniques as they are in the original manuscript, even though one can suspect some kind of paginating accidents in the distant past. The exception is that for ease of reading, we have placed the part dealing with the guards (*leger*) just before the first techniques instead of later as in the original manuscript. Each technique is shown in illustrations with our comments in italics next to them. The translation and the original text are at the top of the page in normal type, our English translation on the left, the original German on the right. We decided to include variations on some techniques (it will be mentioned in the commentary if it's a variation). This is intended to counter the much-too-common misconception that a technique is a set of movements instead of an illustration of a concept. Most techniques can be done in many ways. At the top of some pages you will find the *merk* verse–a rhyme or short sentence–translated above the English main translation; these are Liechtenauer's original rhymes. In the original text, Ringeck included them as introductions to the techniques, but sometimes he added his own short comments. We have included both without separating them. You can thus see both the Liechtenauer verse and Ringeck's own caption.

THE ILLUSTRATIONS

BELOW EACH PICTURE there are black and white "footprints" indicating how you could or should step in a certain technique. Again, do not follow this slavishly; Ringeck says nothing on footwork, perhaps on purpose, so you should experiment. In each picture there are also arrows indicating the movement of a blade and the movement of the feet as well (we can't have too much of that). Also note that some opening illustrations are the same in several places due to the fact that Ringeck describes the same kind of opening or entry.

THE WEAPONS

THIS AREA IS often ignored. Ringeck's techniques work wonderfully with accurate steel replica swords, acceptably with wasters (wooden swords), and very badly with sticks. Draw your own conclusions from this. Try to practice with reproduction weapons for learning the movements and leave the wasters until you are ready to do some two-person mock combat. You should then do some light free-play or sparring with steel blades.

THE INTERPRETATION OF THE TECHNIQUES

WHEN WE BEGAN to work with the material, we were aware from earlier experiences that there is a huge difference in what works when a technique is done slowly versus with full speed and intent. So we decided that each movement must be tested using steel replicas at full speed with the intent to hit. As we began the analysis, one realization came quickly–many of the illustrations in later manuals do not, indeed cannot, depict the intended body posture and alignment when fighting at speed. Counterstriking with the edge proved to work very well as advised by Ringeck, as did voiding using the *indes* principle. We could verify that when fighting in earnest, you counterstrike when you wish to close in, otherwise you will try to void and counter. Also the *absetzen* of the thrusts worked very well. So while we want to stress that this is our interpretation of Sigmund Ringeck's text, we can honestly say that what we demonstrate here we have tried at speed with blunt steel blades with the intent to make actual contact. Again we must emphasize that when done at speed, the techniques will and must be done differently than when done in slow work. And any notions of rigid at-angle blocking will turn out to be impossible.

The text on the left is our translation of the verse and Ringeck's comments.

Our commentaries are in italics next to the illustrations.

The text on the right is Liechtenauer's original verse, followed by Ringeck's interpretation of Liechtenauer's verse in the original German.

Note: The distance between the feet and inclination of the body in some pictures is closer and slightly more upright than what would be the case when you do the technique at speed. It is important to understand that we all will move differently, but that an increase in pace is both likely and recommended when fencing at speed.

Ringeck's Book

Here begins the exposition on the knightly art of the longsword, rhymed by the great master Johannes Liechtenauer. He wrote down the teachings in secret and hidden words so that the art would not become known by all. And these secret and hidden words have been written down and interpreted by master Sigmund Ringeck, the present fencing master to Albrecht, Count of the Rhein and the Duke of Bavaria. He that knows how to fence will now understand these teachings.

This is the teaching. Young knights learn to love God and honor women. Be chivalrous and learn the art that your honor will increase in war. Wrestle well, skillfully wield spear, sword, and dagger in a manly way. Strike true and hard and rush in–those who understand will despise the one who defends. And this you must understand, that all arts have both length and measure.

Hie hept sich an die vßlegung der zedel. In der geschriben stett die ritterlich kunst des langen schwerts die gedicht und gemacht hat Johannes Lictenawer der ain grosse maister in der kunst gewesen ist dem gott genedig sy der hatt die zedel laußen schryben mitt verborgen vnd verdeckte worten darumb das die kunst nitt gemain solt werde Vnd die selbigen verborgenen vnd verdeckte wort hatt maister Sigmund ain Ringeck der zyt des hochgeborne fürsten und herrenn hern Aulbrecht pfalzgrauen by Rin vnd herzog in bayern schirmaister glosieret vnd außgelegt alß hie in disem biechlin her nach geschryben stät das sy ain yeder fechter wol verömen vnd vestan mag der da anderst fechten kan.

Die vor red der zedel. Jungk ritter lere got lieb haben fröwen ia ere so wöchse dein ere kunst die dich ziert in kriegen zu ern hoffiert Ringes gutt fesset glefen sper schwert und messer manlich bederben Haw drin hart dar Rausch hin triff oder las farn daß in die wysen hassen den man sicht brysen daruff dich fasse alle kunst haben lenge und masse.

"If you wish to display skill, stand with the left and strike from the right with it as you step, and left with right leading is how you fence with strength."

Note: This is the first tenet of the longsword: learn to strike properly from both sides so that you learn how to fence well and with strength.

Note: If you want to strike from the right side, make sure that your left foot is forward, if you want to strike from the left, then the right foot is forward. If you then strike an *oberhaw* from the right side, follow the strike with your right foot. If you don't, the strike will be false and ineffective, since your right side will remain behind. Thereby the strike will be too short and not be able to follow its proper path to the left side.

Also: If you strike from the left side and do not follow through with the strike, that is also false. Thus, from the side you strike, follow after the strike with the same foot. Then you will succeed in all techniques. Just so shall you perform all strikes.

"Willtu kunst schowen sich link gen und recht mitt hawen und lick mitt rechten ist das du starck gerst fechten."

Merck das ist die erst lere des langes schwertz das du die hew von bayden sytten recht solt lernen hawen ist das du annders starck und gerecht fechten wilt.

Das vernym allso wenn du wilt howen von der rechten sytten so sich das dein lincker fuß vor stee Und wenn du wilt howen von der lincken sytten so sich das dein rechter fuss vor stee Häust dann den ober haw von der rechten sytten so folg dem haw nach mitt dem rechten fuß tust du das nicht so ist der how falsh und ungerecht wann dein rechte sytten pleibpt dahinden Darum ist der haw zu kurtz und mag sein rechten gang undersich zu der rechten andere sytten vor dem lincken fuß nicht gehaben.

Des glychen wenn du hawst von der lincken sytten und dem haw nicht nachfolgest mitt dem lincken fuß so ist der haw och falsch Darum so merck von welcher sytten du haust das du mitt dem selbigen fuß dem haw nachfolgest so magstu mitt sterck alle dein stuck gerecht trybenn Unnd also sullen alle andere haw och gehawen werden.

"He who goes after strikes has little joy of this art, to close in and strike is your wish, no change comes within your shield, zecke *to the head and body and shun nothing. To fence with your whole body and with force is your wish."*

When you are in the *zufechten* close with your opponent, do not try to avoid his strikes and do not wait for what he intends to do against you. Swordsmen who only wait for the opponent's strikes and do nothing other than avoid them have little success. They are usually defeated.

Note: Always fence with all your strength! Strike against his head and against the body so that he can't change through (*durchwechseln*) in front of your (sword) point. After the strike, from the bind, strike light blows at his next opening, as is described in the following on the five strikes and the other techniques.

In the verse at the top of the page, Ringeck refers to the fact that it is unwise to rely on striking after your opponent has attacked. By "shield" he means the sword as a protection. The word zecke *comes from* zucken *and means to twitch or jerk. He then alludes to the fact that you must strike your opponent hard to wound or kill him.*

"Wer nach gat hawen der darff sich kunst wenig fröwen Haw nachent waß du wilt kain wechsel kumpt in dein schilt Zu koppff zu lybe die zeck nicht vermyde Mitt ganzem lyb ficht waß du starck gerst zu tryben."

Wenn du mitt dem zufechten zu im kumpst so solt du uff sein hew nicht sechen noch warten wie er die gegen dir trybt wann alle fechter die do sechen vnd warten vff aines anderen hew vnnd wellend anderß nichten thon dann versetzen die durffen sich söllicher kunst wenig fröwen wann sy werden do by offt geschlagena.

Item, du solst mercken alles das du fechten wilt das trüb mitt ganzer störck deines lybs Vnnd haw im do mitt nahent ein zu kopff vnd zu lyb so mag er vor dinem ort nicht durch wechseln Vnd mitt dem haw solt du in den anbinden des schwertes der zeckrure nicht vermyden zu der nächsten blöß di dir hernach in den fünff hewen vnd in anderen stucken vßgericht werden.

"Another teaching: hear what is bad, do not fence from above left if you are right-handed, and if you are left-handed do not go from the right."

Note: This lesson describes the methods for left- and right-handed swordsmen. If you are right-handed and have closed with your opponent and believe that you can hit him, then do not strike the first cut from the left side; there you are weak and cannot hold firm if he binds strongly against you. Therefore strike from the right side, then you can work strongly with the sword and use all the techniques you want. Also, if you are left-handed, do not strike from your right side because it is very unusual that a left-handed person can cut strongly from the right side and vice versa.

"Aber ain lere: höre waß da schlecht ist ficht nitt oben linck so du recht bist vnd ob du linck bist im rechten auch ser hinckest."

Mörck die lere trifft an zwu personen ainen lincken vnd ain gerechten Das vernym also wann du mitt zu fechten zu im kumpst bistdu dann gerecht vnd mainest den man zu schlachen so haw den ersten haw nicht von der lingen sytten Wann der ist schwach vnd magst damitt nicht wider gehalten wann man dir starck daruff bindt darum so haw der rechten sytten so magst du starck am schwert mitt kunst arbayten waß du wilt Des gelichen bist du linck so haw och nitt von der rechtten wenn die kunst ist gar wild ainem lincken ze triben von der rechten sytten des gleich ist es och ainem rechten von der lincken sytten.

"'Before' and 'after' are the foundations of the art, 'weak' and 'strong' 'simultaneously' note that word. So you will learn to skillfully work. But if you frighten easily you should not learn to fence."

1) Note: Under all circumstances you must understand the principles of before and after (*vor* and *nach*) since they constitute the foundation of the art of fencing.

Before means preempting your opponent with a strike or a thrust to an opening. Then he must defend or parry (*versetzen*). Be flexible in your parry/defense and aim your sword against one opening after another so that he cannot get through with any of his own techniques. But if he closes in and avoids your sword (*einläuft*), then start wrestling (*ringen*).

2) Note: When you do not succeed with before, then wait for the after. These are the defenses against all techniques he uses against you. So when you must displace him, make it simultaneous (*indes*) and strike immediately from the bind at his next opening. Thus you win the before and he remains in the after. You must also simultaneously note how you can work on the strong or weak [part] of his sword.

This means that from the crossguard of the sword to the middle of the blade is the strong [part] of the sword. With that you can hold firm when someone binds at your sword. The weak [part] is from the middle of the blade to the point, and with that you cannot hold firm. When you understand this well you will be able to work artfully and defend yourself. And lords and gentlemen learn how to survive in earnest and in play by this art. But if you scare easily you should not learn how to fence, for a fearful heart is always defeated.

"'Vor' vnd 'nach' die zway dinck synd aller kunst ain vrsprig 'schwöch' und 'störck' 'in des' daß wort do mitt mörck So magst du lernen mitt kunst arbaiten vnd weren Erschrickstu gern kain fechten nimmer gelern."

1) Merck das ist das du vor allen sachen wol solt verston daß "vor" vnd daß "nach" wann die zway ding sind ain vrsprung do alle kunst des fechtens außgät.

Daß vernym also daß "vor" daß ist das du all weg solt vorckumen mitt ainem haw oder mitt ainem stich im zu der blöß ee wann er dir zu der deinen so muß er dir versetzen So arbayt in der versatzung behentlich für dich mitt dem schwert von ainen blöß zu der anderen so mag er vor deiner arbayt zu seinen stucken nicht kommen Aber laufft er dir eynn so komme vor mitt dem ringen.

2) Mörck magst zu dem "vor" nitt kommen so wart uff das "nach" Das synd die brüch uff alle stuck die er uff dich trybt Das vernym also wann er vorkumpt dass du im versetzen must so arbeit mitt der versätzung "indes" behentlich fur dich zu der nächsten blöss so triffest du in ee wann er sein stuck verbringtt Also gewinstu aber das "vor" und er blypt "nach" Auch soltu in dem "vor" und "nach" mörcken wie du mitt dem wort "in des" arbaitten solt nach der "schwech" und nach der "störck" seines schwertss.

Und das vernym also von dem gehultz des schwertes biss in die mitten der clingen hatt das schwert sein störcke dar mitt du wol magst wider gehalten wann man dir dar an bindt Und fürbass von der mitt biss an der ort hat es seinschwöch da magstu mitt kunst wol arbeitten und dich darmitt wören und fürfass lernen fürsten und herren das sy mitt der selbigen kunst wol mügen besten in schinpff und in ernst Aber erschreckstu gern so saltu die kunst des fechtens nitt lernen Wann ain blöds verzags hertz das tut kain gut wann es wirt by aller kunst geschlagen.

Five strikes learn from the right hand against the opposition. Then we promise that in the art you will be rewarded.

3) Note that this lesson speaks of five secret strikes of which the masters will not know to speak. You should learn no other strikes, when he comes at you from his right side. And try to hit the man with the first strike using one of the five strikes. Whoever can counter you without injury is praised by the master of the teaching as being more skilled than the fencer with no knowledge of the five strikes. How to strike the five strikes is described in the following.

4) Note: Here are named the correct main techniques of the longsword, named in such a manner that you will understand them better. They are seventeen in number and begin with the five strikes.

Note: The first strike is the strike of wrath, *zornhaw*;
the second is the crooked strike, *krumphaw*;
the third is the transverse or cross-strike, *zwerchaw*;
the fourth is the squinting strike, *schielhaw*;
the fifth is the parting or scalp strike, *schaytelhaw*;
the sixth is the four guards, *huten*;
the seventh the four deflections, *versetzen*;
the eighth the traveling after, *nachrayßen*;
the ninth the running over, *überlauffen*;
the tenth the setting aside, *absetzen*;
the eleventh the changing through, *durchwechseln*;
the twelfth the jerk or pull, *zucken*;
the thirteenth the running through, *durchlauffen*;
the fourteenth the cutting of, *abschneiden*;
the fifteenth the pressing of the hands, *hend trucken*;
the sixteenth the hanging, *die hengen*;
the seventeenth the turning, *winden*.

And how you will use the *winden* and the *hengen* and all the other techniques you will find described in the following.

Fünff hew lere von der rechten hand wer das wäre dem wir geloben in kunsten gern zu lonen

3) Merck die zedel setzt fünff verborgene hew da von vil maister des schwertes nicht wissen sagen Die soltu anders nicht lernen hawen wann von der rechten sytten gegen dem der sich gegen dir stöllet zuder were Und versuch öb du mitt ainem haw us den fünffen den man mittdem ersten schlag mügest treffen Wer dir die brechenn kan on seinen schaden so wirt im gelopt von dem maister der zedeln dass im siner kunst baas gelonet soll werden dann ainem andern fechter der wyder die funff hew nicht fechten kann Und wie du die fünff hew howen solt das fündest du in den selbigen funff hewen her nach geschriben.

4) Mörck hie werden genampt die rechten haüptstucke der kunst dess langen schwerts wie yettlichs besunder haist mit dem namen dass du die dester bass versten kündest Der ist sibenzechen an der zal und heben sich an den funff hewen an

Item nun mörck: der erst haw haist der zorn haw der ander der krump haw der drit der zwerhaw der viert der schill haw der funfft der schaittel haw der sechst dass sind die veir hütten das sybent die vier versetzen das acht die nachreisen die neindt die überlauffen das zehend die absetzen dassaylfft das vierzehend die abschnyden das funffzehenddie hend truckendas sechzehend die hengen das sibenzehend das sind die winden Unnd wie du dich mitt den hengen und winden enplösen solt und wie du alle vorgenampte stuck tryben solt das vindestu alles her nach geschriben.

The Sword

When picking up a sword we must understand the anatomy of the sword. Ringeck is very particular, at times, about discerning what part of the sword you should work with in order to get the best result.

A sword, regardless of type, can simply be divided into a strong upper part near the crossguard, and a weaker part reaching from the middle to the tip of the sword. The reason you are stronger in the upper part of the blade (from just below the crossguard to the middle) is that it is closer to your hands and therefore to your body; for the same reason, when you carry something heavy, you keep it close to your body for support. It follows that your weakness at the lower part of the blade (from the middle to the point) stems from its distance from your hands and your body. Therefore you regain a lot of your strength when you fence close to your body.

This is also one of the simple mechanics behind the halfsword techniques. You regain the stability needed in a thrust by extending your grip considerably. By gripping the pommel you will find that it becomes much easier to shift the point by simply using the handle as a leverage. Move the pommel and keep your front hand relatively still, and you will be able to move the sword point much more quickly than by swinging your arms around.

Another important distinction is between the long and the short edge, the long edge being the one on the same side as your knuckles, the short one the opposite side. When you hold a sword, you place your leading hand close to the crossguard and grip firmly, and the rear hand can move as you fence. But you should have a firm grip as you strike or thrust. If you do not, the sword will fly out of your hand if you happen to strike something. (And that is the general idea, to strike something.)

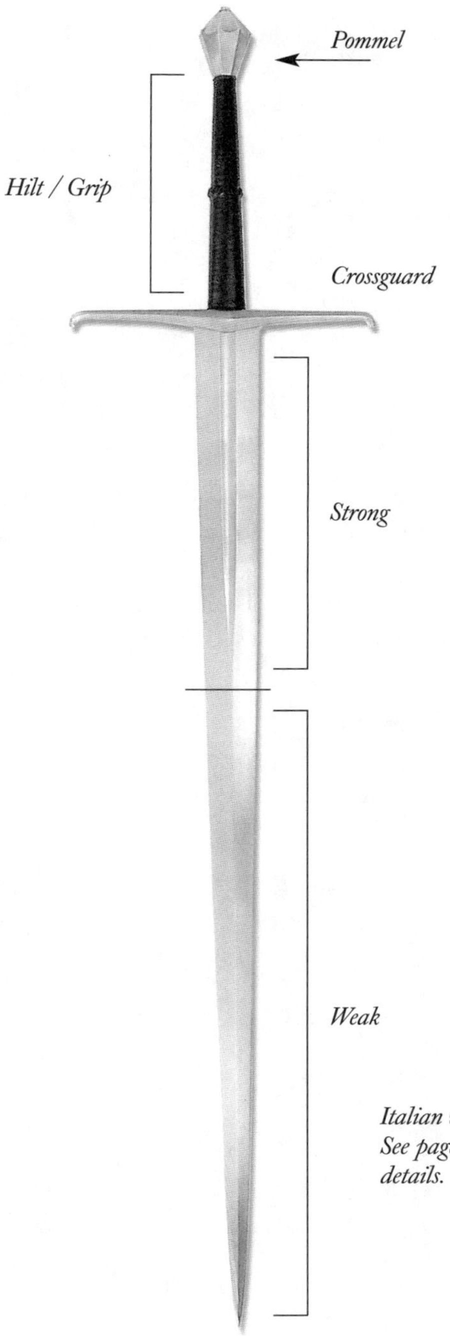

Italian spadona *1450. See page 217 for more details.*

The Techniques

THE FIGHTING TECHNIQUES of Sigmund Ringeck's manual are quite simple and effective. As you read through them, you will find that at times the techniques may seem a bit artificial and even complicated, but remember that Ringeck's book is more than a simple how-to manual; it is a directory of principles of movement that can be used when fencing with the longsword. Since no situation ever occurs twice in combat, learning a set pattern of techniques is of little or no value. What is essential is practicing the possible and most effective principles in such a way to make you familiar with how they work in different situations. So try to read the text and analyze the pictures with this in mind–maximum effect with minimum effort and as little risk as possible.

REGARDING THE TEXT there are several oddities that may puzzle the reader unfamiliar with medieval arms and close combat. Notice that there is no mention of how to fence against a sword and shield. Some techniques invite you to cut, so obviously the opponent does not wear armor. At other times you willingly place the opponent's blade close to you in order to score a hit; either you have armor on or the swords do not have very sharp edges, which is contradicted by the fact that many cuts are shown, and a cut needs a sharp edge. Strikes would depend less on a sharp edge because, without armor, the dull edge would cause severe blunt-force and tissue trauma to the body. Ringeck included techniques or principles that would work both in and out of armor, but without specifying to which category a specific technique pertained, and that, with modification, all techniques may be adapted to armored or non-armored situations.

THE CORE OF Ringeck's and (thus Liechtenauer's) system is condensed and very uncomplicated. In my opinion it consists of the following parts.

THE FIVE STRIKES: These constitute the basics of both attack and defense. Most situations in the book imply that you initiate them from a high position (*krumphaw, zwerchaw, schaytelhaw, schillhaw,* and *zornhaw*).

THE FOUR GUARDS: These positions offer complete coverage of the possible angles you can fence from, as well as the positions where you can end up after completing one of the five strikes. Thus you tend to strike from one guard into another. And since each gives you different opportunities, the fencing will change in a rhythmic and fluid way (*alber, ochs, vom tag, pflug*).

WINDEN AND HENGEN: The turning of the sword. This is the fundamental tool that is used to open a new angle of attack without stepping. *Winden* will let you meet his force with yielding while applying your strength to his weakness to defeat him. *Winden* aims to exploit the liability of force and stiffness; only *indes* will defeat *winden,* and likely by using another *winden* to counter.

DUPLIEREN AND MUTIEREN: These are special cases, such as *duplieren* of a strike, that follow at once from another without the necessity of movement while placing you between his sword and body, an impossible position to defend. *Mutieren,* on the other hand, is a special kind of *winden* where you cross over his sword instead of turning against it.

RINGEN: This can be summarized as a few basic principles of how to trip an opponent or displace his body using your own.

This is the entire core of the system; the rest is window dressing and quite unimportant from the aspect of fighting with a longsword. Sigmund Ringeck must have been a man of practical orientation if he was the fencing master to a powerful nobleman.

With this basic curriculum you are very well equipped for most situations when fencing with a longsword. The aim of every serious combative system, as opposed to those that are more sport, must revolve around simplicity because the stakes are so high. Simplicity is efficiency's best friend, someone has said, and that is very true. The more complex a system is in the sense that it takes time, effort, and talent to master, the less likely it is that it will work in real-life conditions. Fewer components mean you have a good chance of learning to do them very well.

Liechtenauer and Ringeck taught a system that was meant to be used to kill other people. We know from sources later in the 15th century that the art of fencing became more like a sport in that the sword was relegated more and more to the civilian arena. This was not bad in any way for the art, which developed and flourished, but it became less combative. When we turn to these early masters, we must remember that they taught a system of killing efficiently with a sword. We do well to remember this and recognize the inherent simplicity of their teachings.

Vier Leger

THE FOUR GUARDS (*vier leger*) form one of the cornerstones of Ringeck's system. In his foreword he states that you should fence only from these four guards and warns specifically against the bad habits of poor fencers using other positions. Later in his text Ringeck does include other guards as well, but these are less important. They are the *nebenhut* and *schrankhut*, and perhaps the *sprechfenster* and *langer ort* would fit in as well, even though they have more of the character of techniques for specific purposes. The secondary guards are special cases that we will deal with later.

THE POINT IS that each guard offers you different options and opens or closes different avenues of attack and defense. The guards also divide the body into sections for striking or thrusting.

THE FOUR GUARDS must be thoroughly understood because, just like the strikes, they are part of the very core of the system. In fighting you will see that you almost make a dance of the guards if the techniques are done properly. You move from one guard into the next via one or more techniques. Of course it is perhaps not always so orderly in a fight, but the guards offer both starting and ending positions. And we must always understand that any attack should be initiated from where the sword is. If you must move it to place yourself for an attack then something is wrong–you are not dancing well enough. Practicing transitions between the guards and then including the *winden* and *hengen* are essential in order to fence well.

VOM TAG (FROM THE ROOF): A position for the upper strikes as well as an excellent position from which to defend well while being able to move swiftly in any direction.

ALBER (THE FOOL): A position that allows you to strike or thrust from below. Also the position in which you end up after a strike from *vom tag*. More static, but easy to defend from, although somewhat more difficult to launch an attack from.

OCHS (THE OX): A position that offers the point as well as a position from which to wind and deflect with the *kron* or a *zwerch*. Its main function is to offer a thrusting option, but you can be quite vulnerable while in this position due to the fact that it is easy to knock your sword out of the way and rush in.

PFLUG (THE PLOUGH): An excellent guard that offers a serious threat to your opponent in the form of thrusts from below as well as the possibility of *absetzen* against strikes and thrusts with relative ease. A position from which it is very easy to do a *durchwechseln* and counter with a thrust under your opponent's attack.

Note: The Vier Leger *section has been moved here because these moves are the basis for many of the teachings in this book. In the original manuscript they were positioned between* "Schaytler" *and* "Vier Versetzen."

"Four guards only hold, and disdain the common. Ox, plough, fool, and from above should not be unknown to you."

You will not assume any guard except the four above mentioned, how to do that is explained here.

"Vier leger allain davon halt und fluch die gemain Ochs pflug / alber vom tag / sy dit nit unmer"

Ist das du von kainem leger nicht halten solt denn alain von den vieren die hie genant worden sind

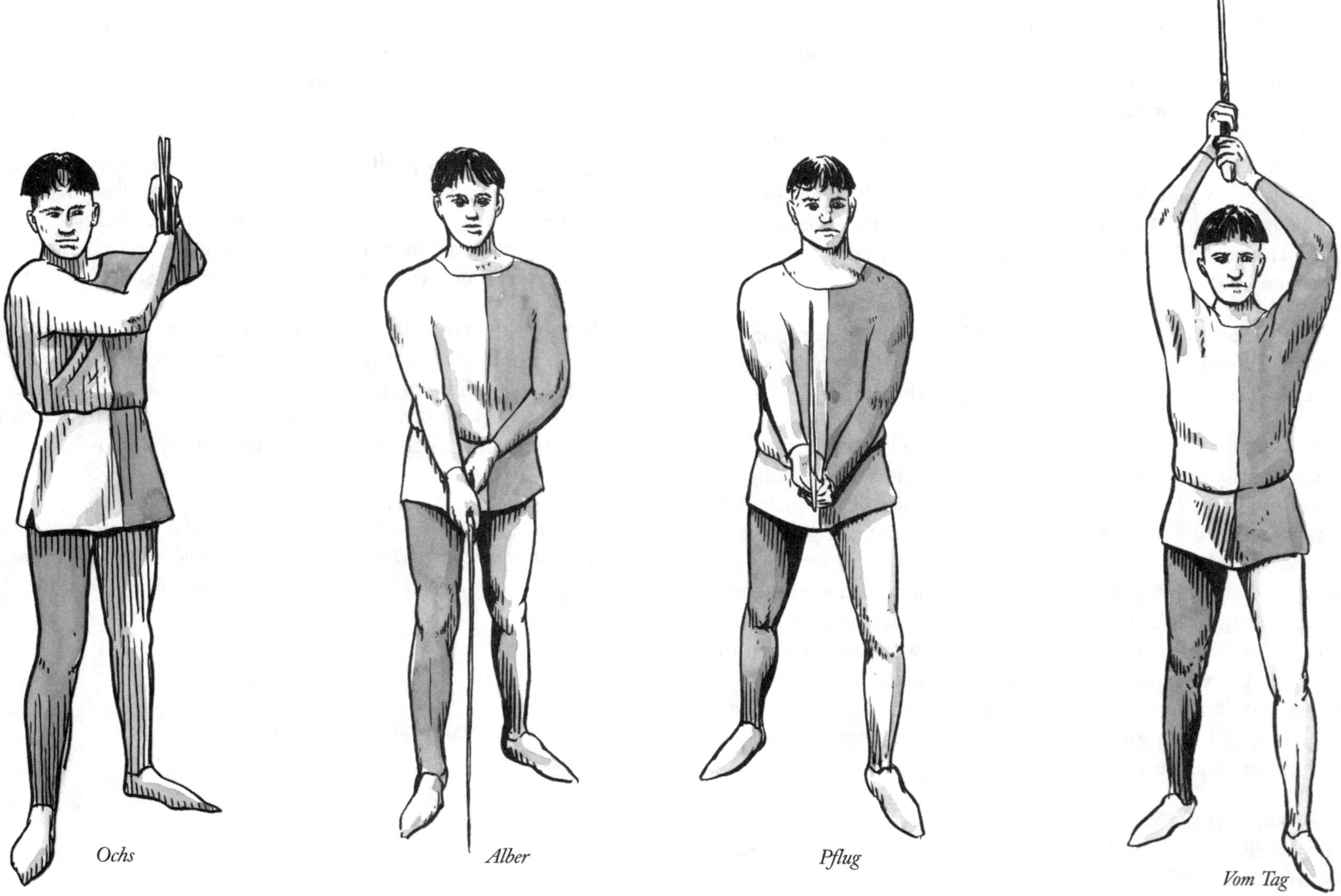

Ochs *Alber* *Pflug* *Vom Tag*

The Fool. Stand with your right foot forward, the sword point toward the ground in front of you with straight arms.

Der alber do schick dich also mit stand mit denem rechten fuß vor und halt din schwert mit uß gerechten armen vor dir mit dem ort uff die erden

Keep your shoulders relaxed and your back straight, with a slight bend in the knees. Do not extend the sword too far out and do not keep it too close. The point should be toward the ground. Naturally you can also do this guard with the left foot leading as shown in the three illustrations below left.

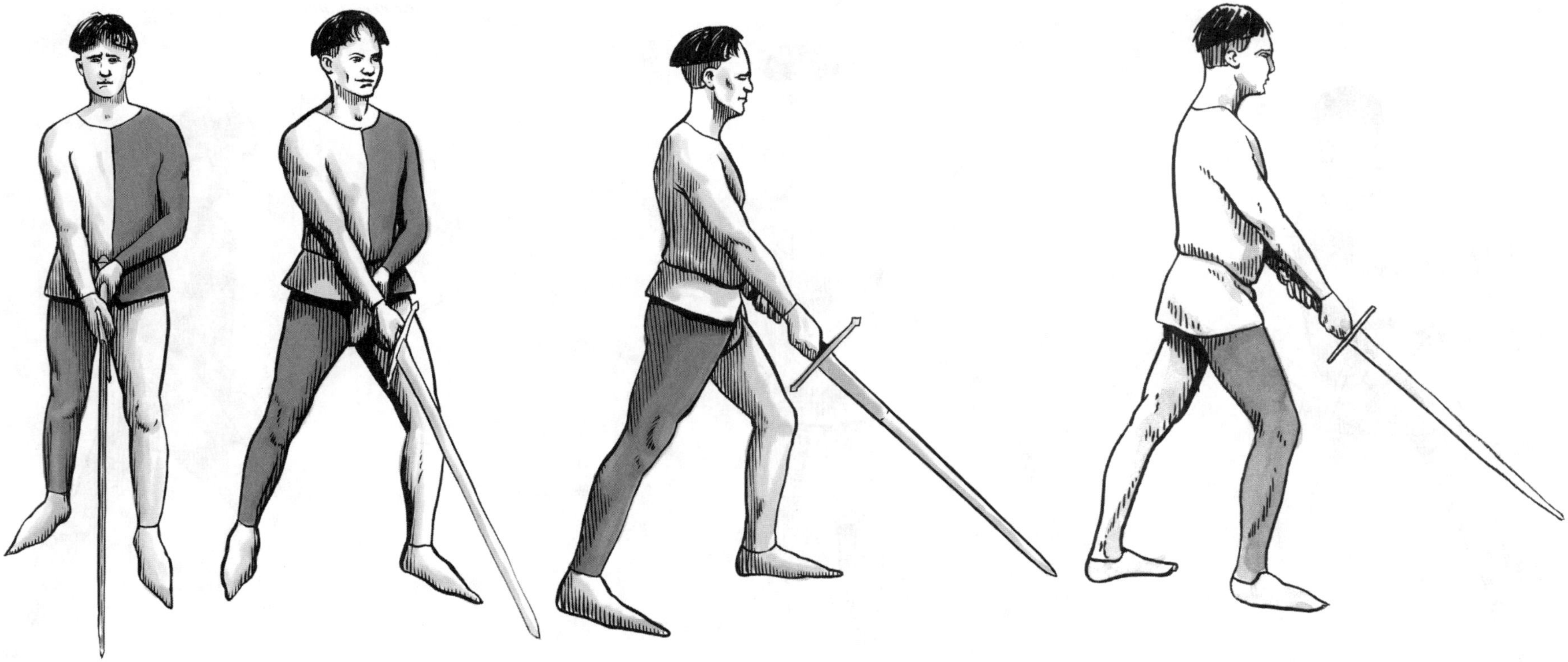

The Plough. Stand with your left foot forward and hold your sword with crossed hands below on your right side, above your knee, with the point aimed at his face.

Der Pflug do schick dich also mit stand mit dem lincken fuß vor und halt din schwert mit gekruetztten henden neben diner rechten sytten ober dienem knye also das im der ort gegen dem gesicht ste

The crossguard should be positioned approximately at your front hip with the pommel below your rear hip. The sword will thus point upward at roughly 45 degrees. Be sure to relax your shoulders, and do not push the blade forward; rather let it hang down, relaxed, as in the illustration below. Your feet should be shoulder width apart. You can hold the position slightly lower or higher depending on the situation.

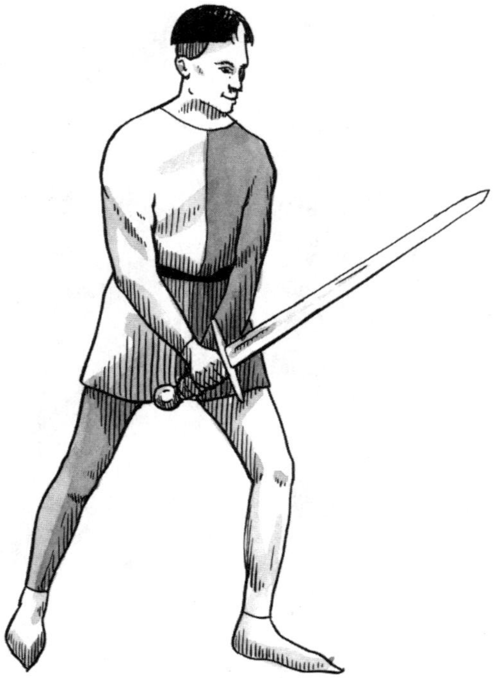

In this position you should again try to aim the point of your sword at your opponent's throat or face. You may turn a bit more or less to the side, but it is important to keep your blade in your centerline as shown below.

The knees should be bent to give flexibility, and the back slightly arched. Do not thrust your chest forward to counter this, since that creates tension that will make you slow. If you are relaxed you can act more quickly. Always avoid tension prior to movement since tensed muscles must relax before they can function properly and make an explosive movement. At the same time, remember that we are all different, and body position always has to be adapted to each individual.

The Ox. Stand with your left foot forward and hold your sword just below your right side of your head. Let the point be pointed toward [your opponent's] face.

Der ochß do schick dich also mit stand mit dem lincken fuß vor und halt din schwert neben diner rechten sytten vir din haupt und laß im den ort gegen dem gesicht hangen

The crossguard should be level with your face. Make sure to explore the difference in holding it close to your head as opposed to far out. Note that if it's too close you might be struck if your opponent hits your blade with force. But if it is held too far out you will become weak and unable to resist. This is often a difficult guard for novices to fence well from.

Keep your weight almost balanced, with just a little more weight on your front foot, and be sure to bend your knees. Try to relax the shoulders even though they are under strain from holding the sword. It is very important not to tense up.

The point of your sword must be aimed at your opponent's face–not because he will be unable to see it (as some claim) but since any threat to the eyes will create a direct response from anyone. Thus threatening the face, even without moving, will be perceived as unpleasant and dangerous by your opponent. The position of the point is also important when considering the techniques that can be performed from this position. The point may be pointed slightly inward instead of straight forward. You can also slope the point slightly down and to the side

From above. Stand with your left foot forward and hold your sword on your right side next to your shoulder. Or hold it with arms stretched above your head. Fencing from these positions will be described in this book.

Vom Tag so schieck dich also / mit stand mit linckem fuß vor und halt din schwert an diner rechten achseln oder halt es mit ußgerechten armen uber din haup und wie du uß der hutten fechten solt / das findest du in dissem buch geschryben

You can also hold the guard above your head as shown below. Make sure that the edge is toward the opponent, and keep the sword back at approximately 45 degrees. If you pull it too far back the crossguard can hit you in the back of your head as you strike.

The crossguard should be next to your chin; do not let it drop below your neck. Keep your shoulders relaxed.

The sword should rest easily in your hands with no tension. It is, however, important that you do not let the sword drop too far down.

Keep your knees slightly bent and relaxed.

Zornhaw

The strike of wrath, or *zornhaw*, is one of the *meisterhaw* in Liechtenauer's system. It is used more frequently than any other strike, and is also the most natural movement for a human being when striking something with an object. It is also the most frequent counterstrike used by Ringeck. The strike has counterparts in all the later systems of swordsmanship that utilize cutting blades up to the use of the infantry saber in the 18th and 19th centuries.

Why it would be considered a *meisterhaw* is slightly strange since there is nothing special about it. But I believe that it is its application as a counterstrike that marks it as unique, since the use of it in offensive attacks is very straightforward. There are also several special follow-ups that seem to be used exclusively, or at least most frequently, with the *zornhaw*, such as the *duplieren* and *mutieren*.

The strike covers diagonally two of the four openings of the body, and it is the strike that is most difficult to defend against with anything except a counterstrike or an evasion. It is also the most powerful strike, since it is very easy to put your body behind it and gain good mass as well as acceleration. And, as Ringeck points out in his basic concepts, it is always advantageous to strike with a step and from your strong side.

"When he strikes from above, zornhaw, [put] *the point in his face."*

Note: If your opponent strikes a blow against you from his upper right side, then strike a *zornhaw* from your right side against his strike. Strike with the strong of your sword and with the long edge. If he is weak at the sword, then thrust with your point along his blade into his face.

"Ver dir ober hawet zorn haw ort im dröwet"

Daß vernym also wann dir ainer von seiner rechten sytten oben ein hawet so haw ainen zorn haw mitt der langen schniden och von diner rechten achseln mitt im starck ein Ist der dann waich am schwert so schuß im den ort für sich lang ein zu dem gesicht Vnnd träw im zu stechen

It is not necessary to step forward with the counter-strike. You can remain still or even step back with your leading foot. What you do will place you at different distances, which in turn will yield different options. But remember Ringeck's advice on stepping while striking, although this is not essential on a counter.

Stand relaxed with the crossguard approximately at your shoulder and the point aimed slightly backward.

As your opponent strikes, strike the blade hard against either the strong or weak portion. This can achieve different effects. If you strike on the weak, the impact will push his blade slightly downward and off to the side. If you strike on the strong part, it will tend to stop the weapon in place. When striking down his strike, the impact will not be edge on edge but rather at an angle of your edge on his flat. This is the counterstrike deflection and the action may even go on to your the opponent in one continuous action.

Maintain weight, momentum, and forward pressure. Lean into the technique, otherwise your opponent will push you back.

In all binds, it is important to distinguish between binding hard or binding weak against your opponent's sword. You must also distinguish between your opponent binding hard or weak against you. A hard bind makes it difficult to execute techniques like a wind and thrust; his forceful bind will push the weak part of your blade aside. A weak bind, on the other hand, allows winden *and other controlling techniques. And remember that it is not possible to listen (feel your opponent's intentions) with the sword in a hard bind; this is also true for your opponent.*

After the strike, lower the point a little and thrust quickly forward. Since you are aiming at his face, it need not be that hard a thrust.

When you move the point of the sword, do this by working with the pommel; do NOT move your arms. That signals intent and is very slow.

Keep your blade in contact at all times so you can feel his movements.

Then take a step forward with your leading foot. This is considerably faster than making a pass with the rear foot and it will also give you greater stability. Use your weight when thrusting. Have a feeling of falling rather than stepping. Thrust by moving the body, not by just extending your arms.

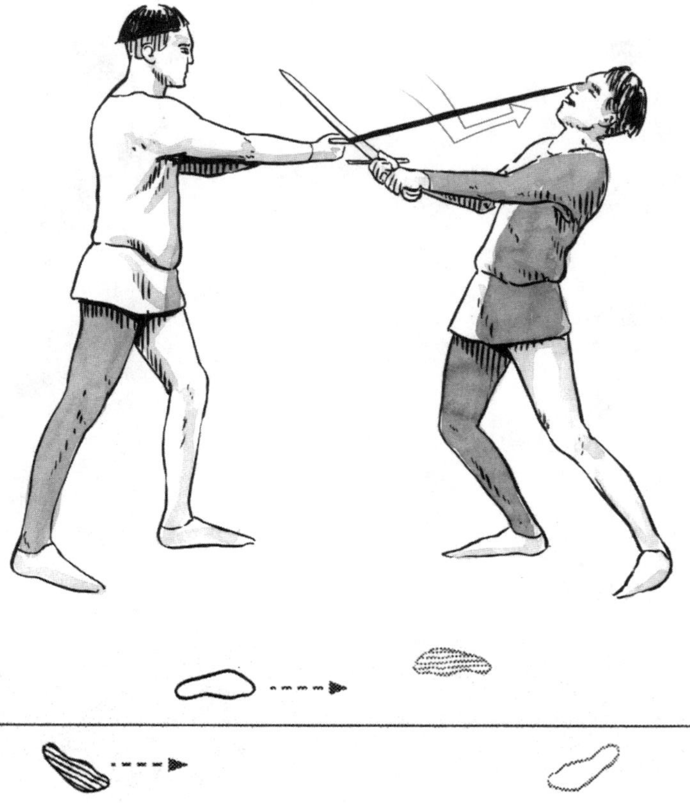

"If he notices, lift it up and move on"

When you thrust from the *zornhaw* at his face and your opponent notices this and deflects (*versetzen*) the thrust with strength, pull your sword upward and away from his. Then strike him from the other side along his sword and hit him in the head.

"Wirt er es gewar so nym oben ab am far"

Wann du mitt dem zorn haw den ort ein schüst wirt er dann deß orts gewar vnd versetzt den stich mitt störcke so ruck dein schwert übersich oben ab von dem sinen Vnd haw im zu der anderen sytten an sinem schwert wider oben ein zuo dem kopffe

After deflecting your opponent's strike, you prepare to thrust at his face.

Since your opponent is trying to deflect your thrust, his force will carry him to the side. Do not resist, but lift your sword and strike at once. Take a small step to the side with either foot to avoid his strike.

Pull your sword up and away and make sure that you are not too close to your opponent, otherwise his sword might strike you. Strike at once at his head. If needed, step or pass back as he strikes and lean forward with your blow.

"Be strong against turn and thrust; if he notices, take it down."

When you strike a *zornhaw* and he deflects it and remains strong at the sword (i.e., maintains pressure on your sword), oppose with strength. Then slide the strength of your sword up to his sword's weak part, turn (*winden*) the hilt of the sword in front of your head, and stab him in the face from above.

"Biß störcken wider wennde stich sich er es so nym es nider"

Wenn du im mitt dem zornhaw jnhawst versetzt er dir daß vnd pleibt dir damitt starck am schwert so byß gen im wider starck am schwert vnd far uff mitt der störck dines schwertes in die schwöch sines schwertes vnd wind am schwert din gehültz vornen für dein haupt vnd so stich in oben ein zu dem gesichte.

Twist your hips and upper body and lift the hilt of your sword to push the opponent's sword away. Keep your blade in contact with his at all times so you can feel his movements.

Do not let your weight and balance go back as you lift your hilt.

Use your weight when you thrust, not your arms; they are too weak. Have a feeling of falling forward rather than stepping. If you are close enough, jab forward with the sword.

Generally speaking it is better to step into a technique with your leading foot instead of making a pass with the rear foot. It is also easier to keep your balance by stepping with your front foot, but the pass will add both reach and power. Either way, always move your feet when acting or reacting. Never stand still. Below you can see that it is possible to step in with the rear foot, but the posture becomes cramped.

"Another technique from the zornhaw.*"*

When you thrust from above (from *winden*) and he deflects the thrust by lifting his hands high and deflecting with his crossguard, remain in the *winden* and thrust downward between his arms and chest.

"Aber ein stuck uß dem zorn haw"

Wann du jm mitt dem winden oben ein stichst alß–vor stett–fört er den hoch vff mitt den henden vnd versetzt mitt dem gehültze den obern stich so plyb also sten in dem winden vnd setz im den ort niden zwischen sinen armen vnd der brust

As you lift to thrust your opponent senses this and responds. You will end up in the ochs *position.*

He raises his sword to the kron *and pushes your blade to the side. The crown guard is a transitory position; you do not fence from this position. The guard is one in which you lift your crossguard above your head and the sword slants forward at roughly 45 degrees down and to the side. From this position it is very easy to strike around and hit him from the left side. Ringeck's text also allows you to lift your crossguard straight up, pointing the sword at the ceiling. At this point Ringeck only states that you must have your hilt high and above your head.*

Remain in contact and push your blade over the outer edge of his sword and thrust down between his arms and chest. As you do this you can take a step forward with either foot as seems best.

You can also thrust below his arms at his breast; the text can be interpreted in both ways. Generally it is a lot faster to thrust like this and it's easier to score a good hit. But you will hit him rather low quite often.

The kron *can also be done with the point of the sword up. In some respects it is better and works well with the techniques.*

"A counter to the abnemen.*"*

When you bind strongly with the sword and he pulls his sword upward and away to strike at your head from the other side of the sword, then bind strongly with your long edge and strike him in the head from above.

"Ain bruch wider daß abnemen"

Mörck wenn du mitt ainem starck am schwert bindest Ruckt er dan sein schwert übersich oben abe von dinem schwert und hawt dir zu der anderen sytten am schwert wider eyn zuo dem kopffe so bind starck mitt der langen schnyden im oben eyn zu dem kopffe

When you bind strongly, you open yourself up to a counter to your other side. Since you commit with force, when your opponent's resistance is removed, you lose balance and time if he moves first. A bind is a quick forceful pressure that maintains contact with his blade.

As he strikes, move your sword over at the same time as you step; use your waist and try to keep the arms still. Use your body and push forward as you displace.

It is not necessary to change feet as you deflect his other strike, but it gives more strength to your sword in the deflecting if you turn your waist.

Strike at once at his head without leaving the bind. Keep your hands high–this way you will strike above his sword and will not need to push it down in order to hit him. If, however, you lower your hands and elbows in the strike, you will have to push his blade down in order to hit him.

Retain balance and forward force. Do not stand up straight, because you then become easy to knock over.

"Note: above; strike, thrust, guards, weak or with strength, indes, vor *and* nach. *Your* krieg *should not be rushed. Who tends to war above, will be shamed below."*

You must immediately notice if he is weak or strong at the sword if he binds against your sword with strike or thrust. When you understand this you will at once (*indes*) know if you should fight him in the before (*nor*) or the after (*nach*). But do not hasten into close combat (*krieg*) since close combat is nothing other than winding at the sword.

"Das öben mörck / haw stich leger waych oder hört in des und vor nach / on hurt dein krieg sich nicht gach weß der krieg riempt oben nyder wirt er beschämpt"

Daß ist das du gar eben mörcken solt wann dir ainer mitt ainem haw oder mitt ainem stich oder sunst an din schwert bindet ob er am schwert waich oder hört ist und wenn du das enpfunden hast so solt du indes wissen welches dir am besten sy ob du mitt dem vor oder mitt dem nach an in hurten solt Aber du solt dir mitt dem an hurten nicht zu gauch lassen syn mitt dem krieg wenn der krieg ist nicht anders dann die winden am schwert

Ringeck talks about the importance of the concept of indes, *which forms a central part of his teaching. Basically it is the ability to react, without a cognitive process, to your opponent's actions. Ringeck shows that you must understand at once–without thinking, for that is too slow–if he is strong or weak at the sword and act accordingly. For example, if he is weak you can push his blade aside with a technique, but he will then be able to feel your movement and intention and respond accordingly. If he is strong, yield and use another approach, using his strength to unbalance him and open him up for a counter.* Indes *is important because it is the instant of opportunity. If you stop to think about it, the moment is gone and your opponent has moved on. To understand* indes *you must not think when you fence, only act and do so with determination. Hesitation is defeat.*

You shall fight like this in close combat (*krieg*): If you strike him with a *zornhaw* and he deflects (*versetzen*) this, lift your arms up high and wind the sword to the upper opening. If he deflects the thrust, remain in the winding and thrust to the lower opening. If he follows your sword and deflects, then direct your point under and past his sword and hang (*hengen*) the point upward and thrust to the other opening on his right side. Thus you can defeat him in close combat, above and below.

Item den krieg tryb also wan du jm mitt dem zornhaw in hawest alß bald er dann versetzt so far wol vff mitt den armen vnd wind im den ort am schwert ein zu der obern blöß versetzt er denn den stich so blyb sten in dem winden vnd stich mitt dem ort die vnder blöß folgt er dann fürbaß mitt der versatzunge dem schwert nach so far mitt dem ort vnder syn schwert durch vnd heng im den ort oben ein zu der anderen blöß siner rechten sytten Also wirt er mitt dem krieg oben vnd vnden beschämpt ist daß du die geført anderst recht kanst tryben

After he has deflected your strike, lift the blade and thrust at his face. When your thrust has been deflected at the upper opening, aim at the lower; when he follows, feel his pressure on your blade.

As you feel his force deflecting you, drop your point quickly under his blade and aim the point at his other side with a winding movement (pulling the sword toward you), then thrust forward quickly.

With this technique it is important not to take steps between the different movements. The thrusts should be done fluidly in a single consecutive movement, perhaps with a small step with your leading foot as you drive the point home.

You can thrust more upward if you want; the pommel will then hang toward the ground and you can thrust at his face/breast on the right side.

"In all winden, *learn to find the strikes and thrusts well and you should try strikes, thrusts, and cuts at every opportunity; then you will be like the master."*

In the winding, always know if you should strike, cut, or thrust. In the winding you should (intuitively) understand which of these three possibilities is the best. Therefore you must not strike when a thrust is better, and not cut when you should strike, and not thrust when you should cut. Be watchful; if one strike is deflected, you hit with another. So if your thrust is deflected, then strike instead. If he rushes at you, then cut from underneath at his arm.

Remember these principles every time you hit or bind the sword if you wish to be like the master who stands before you.

"In allen winden / hew stich recht lern finden auch soltu mit brüffen hew stich oder schnitt in allen treffen den maistern wilt du sy effen"

Daß ist daß du in allen winden hew stich und schnitt recht finden solt also wenn du windest das du da mitt zu handt solt brüffen weches dir under den dreyen daß best sy zu triben also das du nicht havest wann du stechen solt und nit schnidest wann du hawen solt und nicht stechest wann du schniden solt Und mörck wan man dir der aynes versetzt das du in mitt dem andern treffest Also versetzt man dir den stich so treyb den haw Laufft man dir eynn so treyb den undern schnitt in sin arm.

Des mörck in allen treffen und anbinden der schwert wilt du anderst die maister effen die sich wider dich setzen.

Here Ringeck is writing about the end product, so to speak, and pointing out that in any given situation many things can be done, but that one solution will be better than others. Again, it is not a thought process that yields the right movement, but an intuitive action that reflects the intent of your opponent. And only one action will be perfectly right at any given time in combat.

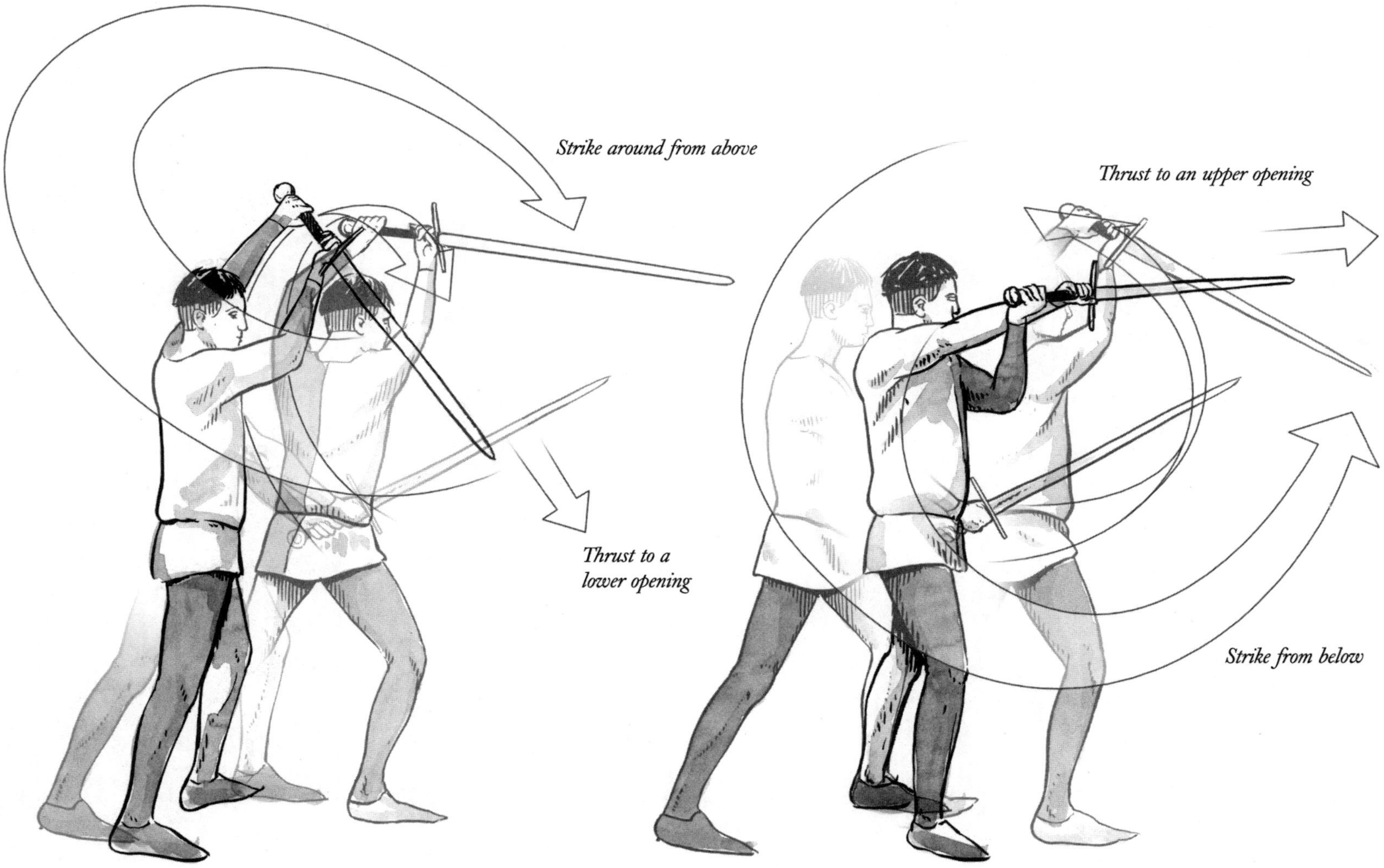

Here are some possible movements that we can do in winden *as described on page 46. It is important to remember that there are many alternatives and that in principle it is possible to execute every movement from* winden.

The Four Openings

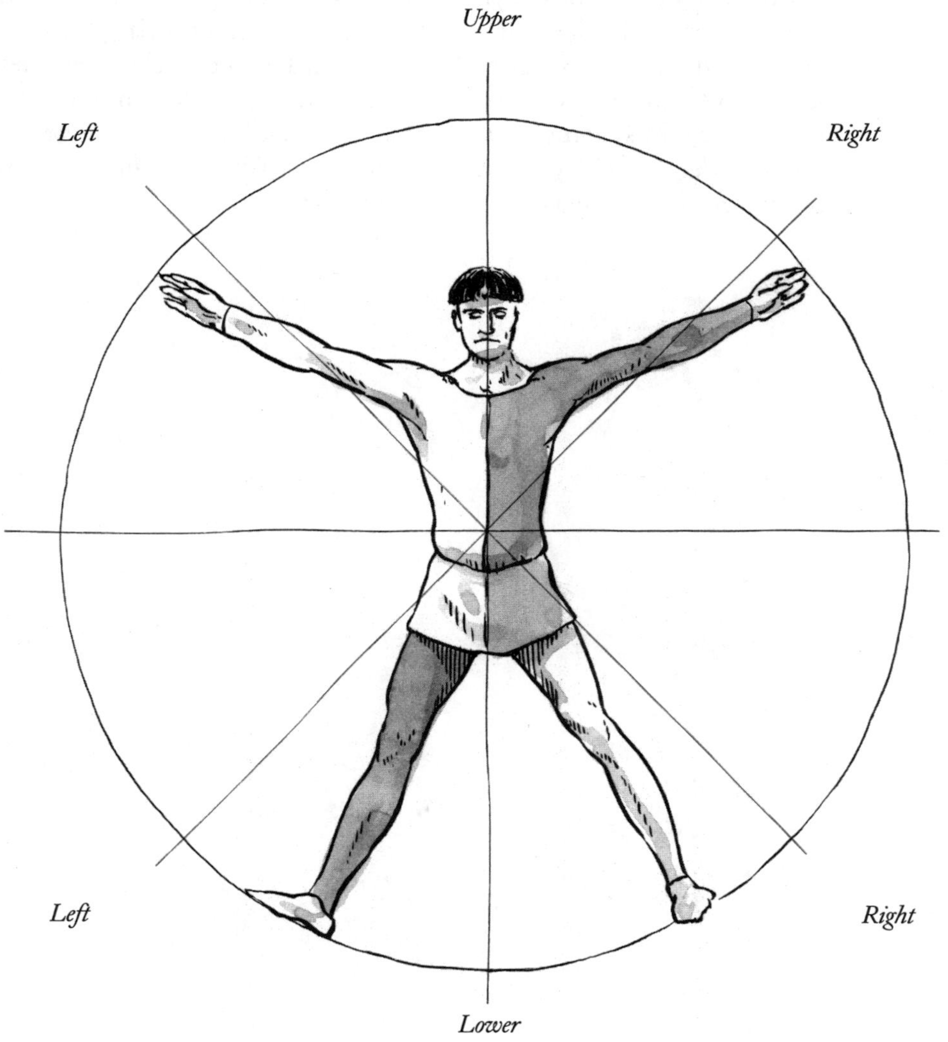

THE IMPORTANT ISSUE here is the understanding that the division of the body into four openings is a handy guide for striking. The vertical and horizontal lines divide the body into two upper and two lower openings.

REMEMBER THAT NO matter how your opponent's sword is placed, one or more of the openings will always be less protected. When you attack, think in terms of these openings when you feint or make a series of attacks, since from certain defensive positions it becomes hard to defend against certain attacks aimed at other openings. With the four openings it is important to note that each guard (*leger* or *huten*) will close or open specific openings and allow for a set number of the nine ways to attack to be used. Without understanding these interrelationships between openings and the ways of attack, you can't understand fencing.

THIS FIGURE ALSO serves to show that there are only eight ways to place a strike on a human body. Of course variations are possible, but a diagonal descending cut can only be made one way (even if it may strike a high or a low target). Apart from the eight strikes, you can also thrust–making nine possible ways to attack one of the four openings. It is a useful tool because it emphasizes the simplicity that tends to be forgotten when we delve into the plethora of recorded techniques.

IN YOUR MIND'S eye, try to place this figure over your opponent so you can see which of the openings are open or closed by the stance he has taken.

"Know the four openings, aim and you will surely hit; pay no heed to what he does."

These are the four openings that you should aim at in combat. The first opening (*bloß*) is on the right side, the next one is on the left side above the belt. The next two are on the right and left side below the belt. When closing, immediately notice which openings are open, then aim. Thrust with the long point (*langen ort*), practice traveling after (*nachrayßen*), and other possibilities. Do not concern yourself with what your opponent is doing: fence securely and you will strike so well so that he will not get through with his own techniques.

"Vier bloß wisse / ram so schlechstu gewisse an alle for / on zwifel wie er gebar"

Hie soltu morcken die vier blossen an dem man da du all wegen zu fechten solt Die erst bloß ist die recht seytt die ander ist die link seytt oberhalben der girtel deß manß Die ander zwuo sind och die recht und die linck seytten underhalben der girtel Der blossen nym eben war in dem zu fechten mitt welcher er sich gegen dir enblösse der selbigen reme küntlichen on far mit einschiessen des langen orts mitt nachreissen und sunst nitt allen geförten und acht nitt wie er mitt synen geförten gegen dir gebar So vichtest du gewiß unnd schlechst schlege daruß die do treffenlich sind und laust in domitt zu seinen stucken nitt komen

Mutieren and Duplieren

Mutieren is a special kind of *winden.* All *winden* work on the principle of capturing his weakness with your strength and the thrust or strike. With the *winden* you can of course wind without blade contact; you are then exploiting a weakness of your opponent since he has left an opening. In *mutieren* you capture his weakness by going over his sword and down in the thrust, which you do not do in the ordinary *winden.* This is also a technique that can only be done from the bind and only when he is weak, or you will be struck again. However, since you are capturing his sword, you are also seizing his attention and putting him at a disadvantage in regard to time. Because it is somewhat risky, *mutieren* must be done with determination and speed to succeed. After the technique you must change through at once to avoid a counter. You will also be in a very good position to close in and grapple him, which will be easy if you have just stabbed him in the gut or inner thighs.

Duplieren is a special kind of counter or follow-up that can be done only from the bind (although the same movement can be executed as a horizontal strike from left to right at any time, but without the advantages of the *duplieren*). In the bind you have the chance to angle your sword in between your opponent's sword and his body. This can only be done on one side from each bind, as you will see if you try it. The technique deserves special attention since it is the fastest and surest way of scoring a good hit in your opponent's face with a minimum of effort. It is what you would call an automatic hit. But the *duplieren* can only be done if your opponent binds weakly. If he binds hard and you do a *duplieren,* you will hit him but you will also be struck in the upper body as his sword is released from the bind. From the *duplieren* it is very easy to do a variety of *winden* actions as well as cuts in the event that your opponent manages to deflect your strike by assuming the *kron* guard. By lifting his crossguard very quickly above his head, he does stand a chance of deflecting the attack.

"If you want to figure out how to artfully break open the four openings, above dupliere, *below right* mutiere. *I tell you honestly a master will not defend without danger. And if you have learned well, he will not come to blows."*

When you wish to break open the four openings artfully, do like this: *duplieren* to the upper opening to the strong of his sword and *mutieren* to the other opening. I tell you that he will not be able to defend himself with either thrusts or strikes.

"Wilt du rechen die vier blossen künstlich brechen oben duplier unden recht mutier ich sag dir für war sich schütz kain maister an far hastu es recht vernommen zu schlage mag er klain kommen"

Daß ist wann du dich an ainem rechen wilt also das du im die vier blossen mitt kunst wilt brechen so tryb das duplieren zu der oberen blössen gen der störck seines schwerts und daß mutiern zu der anderen blöß So sag ich dir für war daß er sich dar vor nittt schützen kan und mag weder zu schlachen noch zu stechen komen

The first part is a simple duplieren*; you can do this after the bind or start from a* krumphaw *and thus initiate the attack. After doing this you can* mutieren *to his other side*

To mutieren *after the strike, wind the blade so that your hands become uncrossed and thrust at his other side, either straight or at a downward angle.*

The duplieren

If you strike a *zornhaw* or another *oberhaw* and he displaces this with strength, then at once (*indes*) thrust the pommel of your sword, using you left hand, under your right arm. With hands crossed strike behind his blade, between his blade and his body, across the face. Or strike him on the head.

Das duplieren

Item wann du in mitt dem zorn haw oder sunst oben eyn hawst versetzt er dir mitt störck so stoß /in des/ deines schwertes knopff under deinen rechten arm mitt der lincken hand und schlach in mitt gecrutzten henden am schwert hinder sines schwertes klingen zwischen dem schwert und dem mann durch daß maul Oder schlach in mitt dem stück uff den kopff

Note the mutieren

Do the *mutieren* like so; if you bind his sword with an *oberhaw* or another technique, wind your short edge at his sword, and move up with the arms and hang the blade of your sword over his sword and thrust him in the lower opening. And try this on both sides.

Mörck daß mutieren

Daß mutieren tryb also wenn du im mitt dem obern haw oder sunst an daß schwert bindest so winde die kurtzen schniden an sin schwert und far wol uff mit den armen und heng im dein schwertes clingen ussen über sein schwert und stich im zu der undern blösse und das tryb zu bayden sytten

As before, wind your sword up and back up a bit so that you reach the weak part of your opponent's sword.
This must be done with speed.

The crucial instant is when you turn over his blade. This must be done with speed and determination. Your thrust begins there. You can't move around his sword and then thrust to his lower opening. It is one movement, not two.

Turn your sword over his (imagine that you are entwining his blade) and control him by capturing the weak part of his blade with your strength. Thrust down along the outer side of his blade to his lower opening. Step forward with your lead foot or make a pass with the rear foot as you see fit.

Remember that the lower opening is below the belt; that is, you can stab him in the lower abdomen or in the groin, not just in the lower leg. In fact, it is considerably easier to place a thrust from mutieren *in the upper part of the lower openings. A hit there is also somewhat more difficult to void or deflect.*

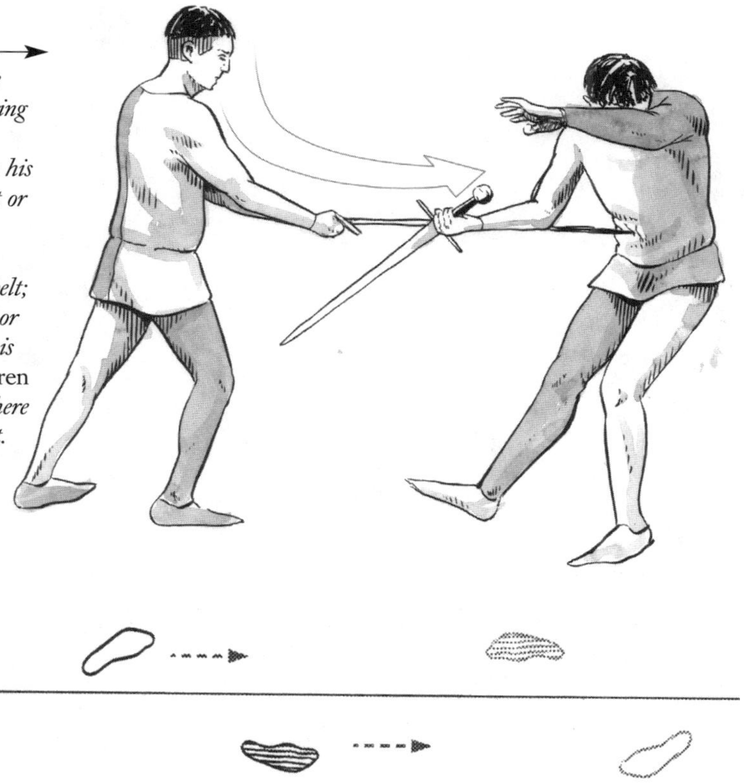

Krumphaw

The krumphaw, or crooked strike, is a very versatile technique and may be employed in defense as well as in offense. The name derives from the fact that your arms are bent, and most often crossed, thus rendering all strikes with crossed, or crooked, arms–a *krumphaw*, in a sense.

The variations of the strike are well suited for breaking up an opponent's defense as well as negating his attacks. The most common position to start from is the upper guard (vom tag), from where it is very easy to position the blade to make a vertical *krumphaw*.

A krumphaw can basically be made from any direction and, as we shall see later, it is a good strike to make from the *schrankhut* (described later). As we will see, the *krumphaw* is seldom used as a way of initiating an attack, but more as a follow-up tool or as a counter to an attack by your opponent. Unfortunately, Ringeck is not specific on how to execute this strike, such as whether the long or the short edge should be used.

Later, in 1570, Joachim Meyer stated that it is the long edge that should be used, although Hans Talhoffer (c. 1440s) shows us a picture indicating the short edge. Well, both work fine, each offering different advantages in the follow-ups.

Krump *means crooked or bent, so a* krumphaw *will be a bent or crooked strike.* Krumphaw *with the short edge.*

"Use the krump *with agility; aim the point at the hands."*

This is how you should aim a *krumphaw* at the hands: when he attacks you from his right with an *ober* or *unterhau* at the opening, leap with your right foot toward his left side and simultaneously strike his hands with your arms crossed, using the tip of your sword. You can also use this if he is in the *ochs.*

"Krump uff behende / wirff den ort uff die hende"

Daß ist wie du krump solt hawen zu den henden und daß stuck tryb also wenn er dir von deiner rechten sytten mitt ainem obern oder undern haw zu der blöss hawet So spring auß dem haw mitt dinem rechten fuß gegen im wol uff sin lincke sytten und schlach in mitt gecrützten armen mitt dem ort uff die hende und das stuck tryb och gen im wenn er gen dir stant in der hüt deß ochsen

Remember to use your left hand to push the pommel under your right arm to execute the strike while you step. Do not step and then strike: both strike and step must finish at the same time for maximum effect.

A krumphaw *with the long edge. Also, your opponent's sword will be pressed down by your strike, or he will most likely drop it. You can strike when he is starting his strike, or change through and aim at his hands as the strike is finishing.*

A strike with the short edge. Either edge works fine. The step out is the crucial moment; the strike is an "automatic hit" so you hardly have to aim! Remember that it is more effective to strike at roughly 45 degrees forward than to strike straight to the side. You catch his strike sooner and can easily make a cut or strike at his face.

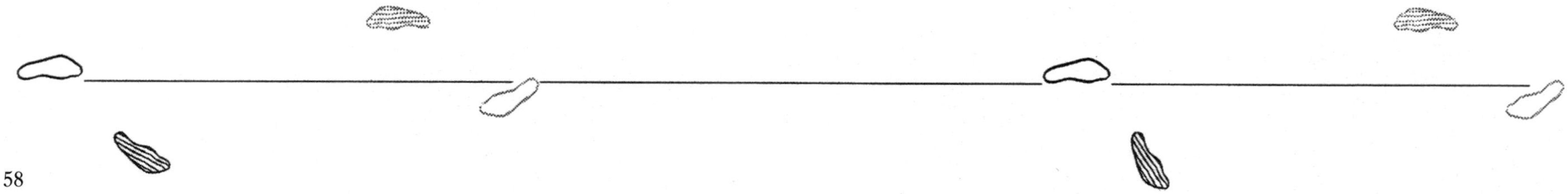

To use the krumphaw *against an* ochs *position, simply step toward your opponent and strike toward his hands or his sword. This does require timing, since you will be stepping toward the tip of his sword, but there is less risk in stepping to his inside than his outside.*

Against an unterhaw, *perform the strike the same way as before. You will, however, need to drop the point of you sword sharply in order to strike your opponent's hands in time. Remember to step toward his open side. Always make sure that the strike, the feet, and the body finish their movements at the same time. Otherwise much of the power in the strike will be lost.*

"Use the krump *and deflect with a step, then he will strike last."*

This is how you can deflect strikes from above using the *krumphaw*. If he strikes at you with an *oberhaw* from his right at your opening, then leap to his left side with your right foot and move the tip of your sword over his sword to a *schrankhut*. Practice this on both sides. From the deflection you can strike at his head.

"Krump wer wal setzet mitt schrytten er wil hew letzet"

Daß ist, wie du mitt dem krumphaw die obern häw absetzen solt daß stuck tryb also Wann er dir von siner rechten sytten oben ein hawet zu der bloß so schryt mitt dem rechten fuß uff syn lincke syten über sin schwert / mitt dem ort uff die erden in die schranckhüte daß tryb zu bayden sytten Och magstu in uß dem absetzen uff das haupt schlachen

Schrankhut *means fence, or barrier guard. A* schrank *is a vertical piece of wood incorporated, for example, in a fence. The name derives perhaps from the vertical position of the sword, which resembles that part of a fence or a fence post.*

As the strike comes, deliver a krumphaw *at his blade and continue the movement until your point is aimed at the ground. Move the tip of your sword over his and continue downward with force, making a half-circle with your sword tip. At completion your sword should be almost vertical, with the point toward the ground.*

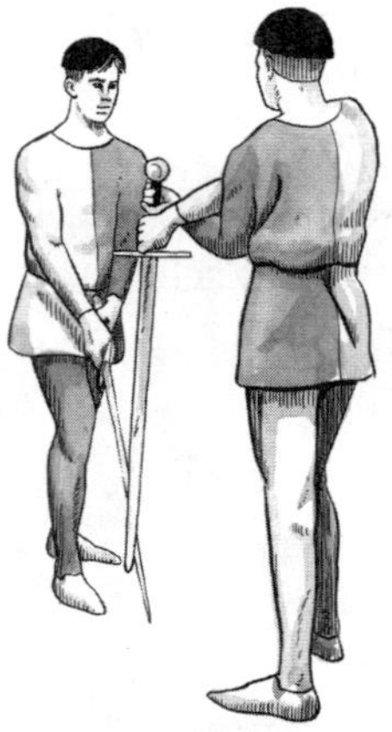

Schrankhut *is seen at left.*

*Since your opponent's sword is moving at an angle to the direction that you are swinging, you will easily be able to strike his sword down. In doing this you will retain the control of his sword, and can feel (*fühlen*) if he tries to do something. "Listen" with your blade.*

From the schrankhut *strike at once upward at his face or throat using the foremost part of your blade in a wide, cutting motion.*

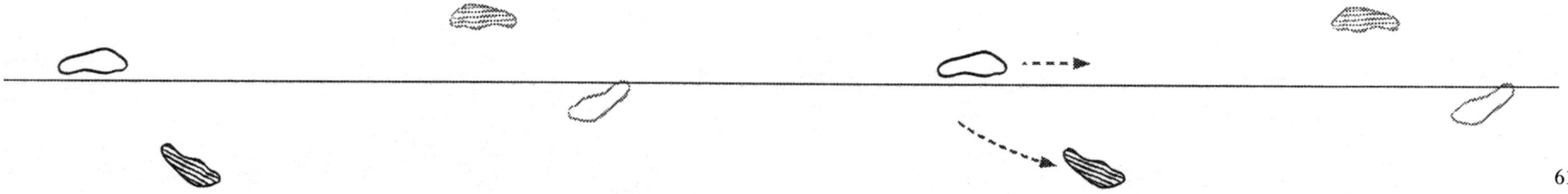

"Use the krump *at the strike and you will weaken the master."*

If you want to weaken a master, then when he strikes an *oberhaw* from his right side, strike a *krumphaw*, with your hands crossed, at his sword.

"Haw krump zu den schlechen den maistern wiltu sy schwechen"

Daß ist wenn du ainen maister schwechen wilt so tryb das stuck also wenn er dir oben einhawt von siner rechten sytten so haw krum mit gekreutzten henden gegen sinen haw uff sin schwert

The idea is to strike the incoming blade from the side, thereby pushing it and your opponent's arms to the side. From there it is easy to make a finishing strike. This requires you to be in balance and push slightly forward with your body weight. The simple idea behind breaking a strike is that your opponent will move in such a way as to have perfect balance and power at the moment of impact, not before. That is the only way to do a powerful and sure strike. But if you by some means were to break his strike abruptly in midair, he would become unbalanced. You, on the other hand, will be in balance and with your power well developed because you calculated the impact at precisely this moment and are therefore prepared for it. Not only krumphaw *but also* zornhaw, mittelhaw, *and* zwerchaw *can be used this way.*

You can either take a step forward or remain still as you strike at his sword. If you step forward you will come quite close and must direct the strike accordingly. You will then be in a very good position for some wrestling. If you remain still, try a thrust with a step with your lead foot to close the distance and to gain force. You can also step back and strike his sword, and then come around in a mollinet*-like maneuver (to strike around in a circular movement where the strike does not stop but continues around completely) to hit his front leg from the outside.*

The krumphaw *can be done with the long edge or, as here, with the short edge.*

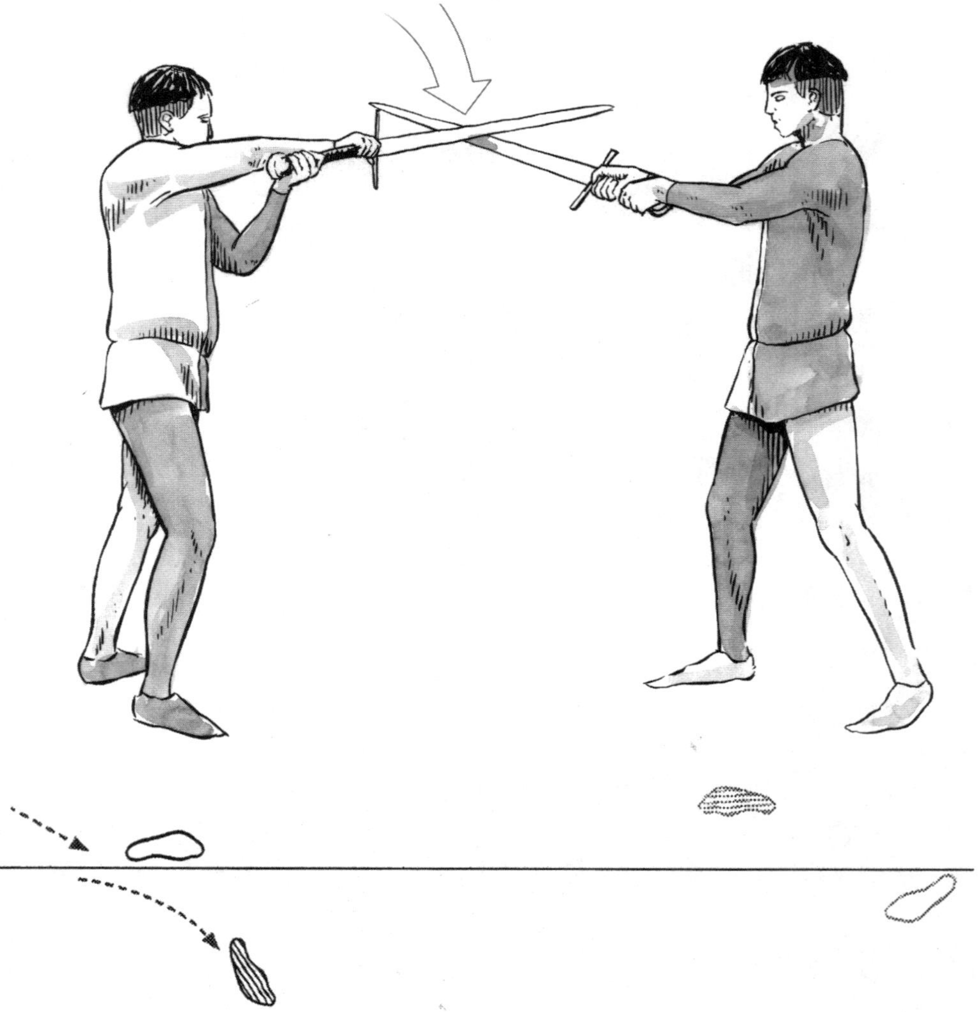

"When he comes from above, stand away; that I will laud."

If you strike him with a *krumphaw* against his sword, then strike at once from the sword using your short edge upward at his head. Or after a *krumphaw,* wind the short edge against his sword and stab him in the chest.

"Wenn es klutzt oben / so stand ab das will ich loben"

Daß ist wenn du im mitt dem krumphaw uff sin schwert hawst so schlache vom schwert mitt der kurtzen schniden bald wider uff / im oben ein zu dem kopff Oder windt im mitt dem krumphaw die kurtzen schnyden an sin schwert und stich im zu der brust.

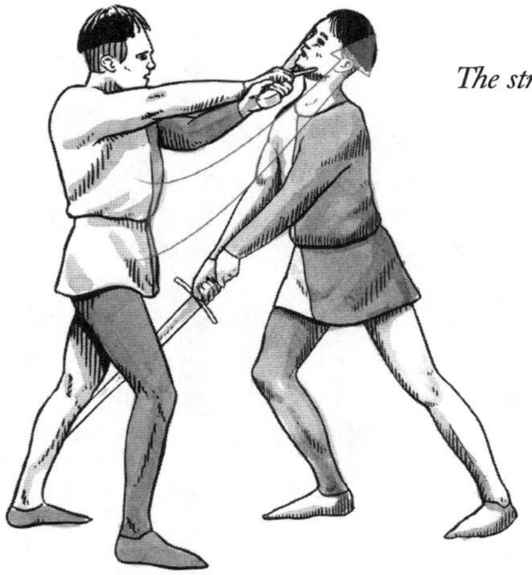

The strike with the back edge is a simple backhand with a lifting motion.

After the bind, strike at once up to his face; this can be done with either the long or short edge of the sword. Take a small step forward with the lead foot.

After the bind, wind the short edge by pulling and turning the sword toward you. Step forward with the lead foot and thrust with force.

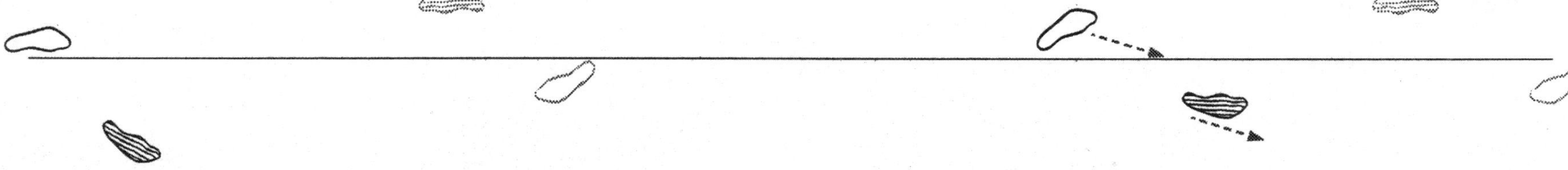

"Do not use the krump, *strike short and change through; that will work."*

If he leads with an *oberhaw* from his right shoulder, proceed as if you want to bind his sword with a *krumphaw.* But let your strike fall short and guide your point through under his sword and wind your hilt over your head on your right. Then thrust at his face.

"Krum nicht kurtz haw / durch wechsel / sol dar mitt schow"

Daß ist wenn er dir von siner rechten achseln oben ein will howen so tu alß ob du im mitt dem krumphaw an sin schwert wöllest binden unnd kurtz und far mitt dem ort under sinem schwert durch und wind uff din rechte sytten dein gehültz über din hoppt und stich im zu dem gesicht.

The trick here is to miss your opponent without making him aware of the feint. Continue the movement and wind the sword without stopping. Then thrust to his face or neck from above.

It is very important that the thrust is one flowing motion, since breaks will give your opponent time to mount a defense. It is one long, smooth movement from the beginning of the feint to the end of the thrust.

You can again take either a step forward with your lead foot as you thrust or make a pass with your rear foot as above. What you do depends on the situation.

As always, you can also strike with the long edge; then the thrust will be on the other side with the arms uncrossed.

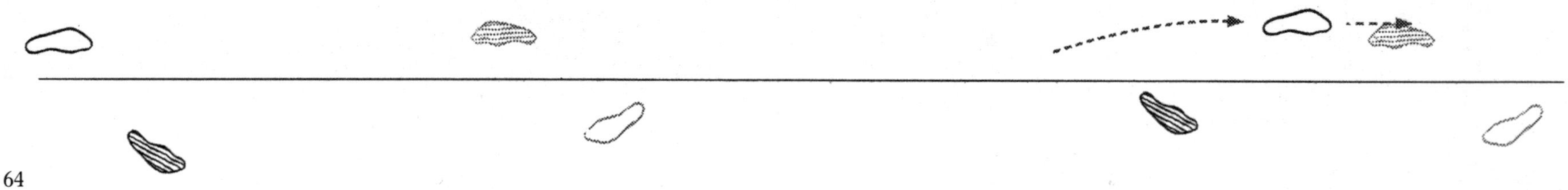

"If he uses the krump *to mislead you, the noble fight will confuse him so that he will honestly not know what to do"*

If you attack him from your right with an *ober* or *unterhaw*, and he deflects the strike with a *krumphaw* with crossed arms from the right, then stand steady with your sword against his. Then thrust your point under his sword at his breast with extended arms.

"Krump wer dich Irret / der edel krieg in verwürret daß er fürwar nicht wayst wo er sy one far"

Daß ist wann du im von rechten sytten ober oder unden zu hawest hawt er dann och von syner rechten sytten mitt gekreutzen armen krump uff din schwert und verirret dir do mitt dein haw so blyb mitt dinem schwert starck an dem sinen Unnd schüß im under dem schwert den ort lang ein zu der brust

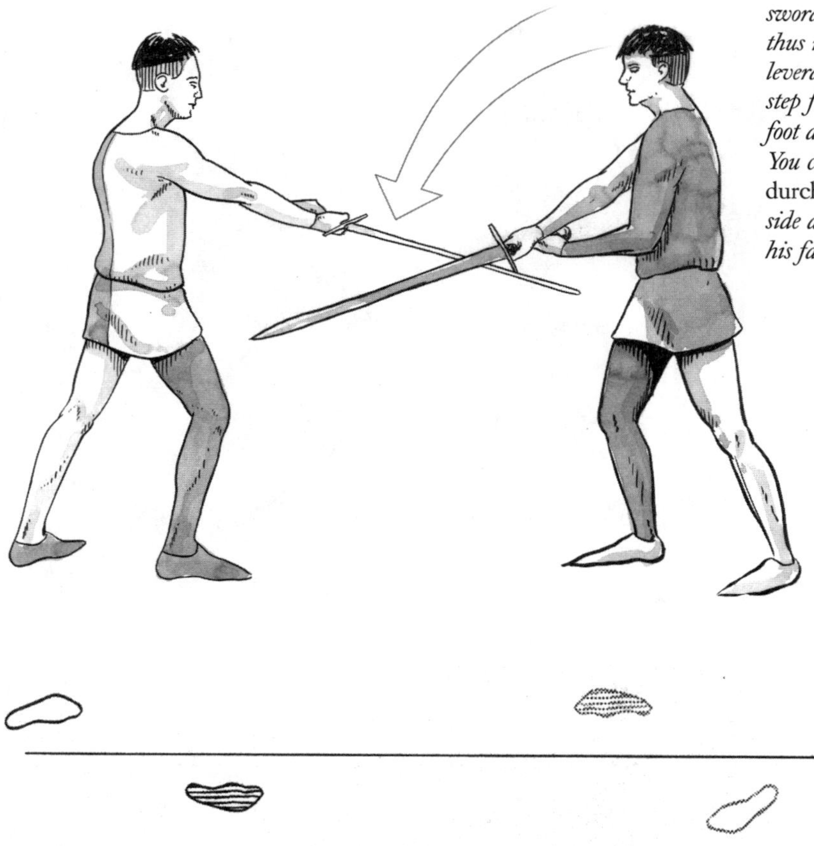

As he presses down on your sword, lower your pommel and thus raise the point using the leverage of the sword. Take a step forward with your lead foot and thrust at his breast. You can also change through durchwechseln *to his open side as you thrust, then aim at his face instead.*

"Another technique against the krumphaw."

Another defense against the *krumphaw*: if you attack him from your right with an *oberhaw*, and he deflects the strike with a *krumphaw* with crossed arms from his right side and pushes your sword toward the ground, then wind to your right side. Move your arms over your head and place the sword point high on his breast.

"Ain andern bruch uber den krumphaw"

Mörck wenn du im von diner rechten sytten oben ein hawst hawt er denn och von siner rechten sytten mitt gekreutzten armen krump uff dein schwert und drückt dir das da mit under sich gen der erden So wind ge deiner rechten syten und far mit den armen wol uff uber dein haupt und setze im dein ort oben an die brust

When he pushes your sword down, offer some resistance and then wind to your right. This will free your sword from his and leave him vulnerable. When you wind, do so by manipulating the crossguard. Then the sword will follow smoothly.

To add force, step forward as you thrust. Move either your lead or rear foot, depending on the situation, and thrust at his face or throat. The lead foot will be faster than the rear foot.

If he deflects this, remain standing with the hilt in front of your head and skillfully work the sword point from one opening to the next. This is called the noble fight, and thus you will confuse him so much that he will no longer know how to defend himself.

Versetzt er dir daß so plyb also sten mit dem gehultz vor dem haupt und arbait behendtlich mit dem ort von ainer bloss zu der anderen Daß hayset der edel krig da mit verwirstü in so gar daß er nit waysst wo er vor dir blibe sol fur war

Here is just an example of how to do this. It can be done in many ways using winden, duplieren, *and* mutieren *in different combinations with* durchwechse*ln to find new angles.*

If he deflects your thrust by raising his crossguard, aim for one of the lower openings.

Do this by lowering the crossguard outward and down (the sword thus turns while moving, allowing your arms to uncross), and in the same movement thrust at his lower opening.

Should he deflect this as well, wind to your right or left and continue to thrust. You can also, after winding, aim a straight kick at his knee (try to hit it from the front or side to break it). Or step closer and perform a drawing cut while passing on either side.

Zwerchaw

The zwerchaw (transverse, or crosswise strike) is one of the *meisterhaw.* The word *zwerch* denotes a movement from one side to the other on a more or less horizontal plane. It is also one of the *versetzen,* meaning that it is one of the strikes that counter an attack. It is said to break all attacks that come from above. It also counters the guard *vom tag,* since the *zwerch* breaks what comes from above. It is an excellent strike with which to initiate attacks as well. Due to the fact that it is very easy to step up to strike on either side with great speed, it is very versatile for taking the initiative and pressing your opponent. The movement itself also lends itself to alternating high and low strikes, a combination of attacks that becomes very difficult to avoid.

When struck on the other side (from your off-hand side), the long edge is used instead of the short because it becomes impossible to strike on your off-hand side with the short edge with any degree of efficiency (you can only do that on your strong side). It is curious that the simple horizontal strike is not mentioned by Ringeck, but that may be due simply to the fact that it is a natural movement and his text deals more with the higher aspects of the art. And the *zwerch* also offers many of the advantages of the horizontal strike, but fewer of its weaknesses.

The important points of the *zwerch* are to step toward the side you wish to strike and to make sure that the crossguard is well over your head (or at least that you are aware of the dangers of lowering it). Naturally, if you strike a low *zwerch* the crossguard must be lower than your head or at least just level with it.

1st position: vom tag　　*2nd position: side*

Zwerch *means diagonal, transverse, or crossing of something (e.g., an imaginary line). So a* zwerchaw *is a transverse or diagonal strike.*

This is a basic overview of how the zwerchaw *is done. On your right-hand side you should strike with the false short edge by turning the crossguard in front of your head so that the knuckles of your right hand are on top when you hit. Step forward and to the side with your right foot as you strike.*

The zwerchaw *is a very useful strike to master because it can be used both to counter your opponent's upper strikes and make a very efficient attack. Its movement is difficult to grasp at first, but the important thing to remember is to strike with the short edge.*
From the position of vom tag, *begin to turn the blade as you begin to step forward (or backward as the situation demands).*

What it might look like from the front. Depending on the situation, your sword might be more in front of you. The important part, however, is that your body is aligned with your target.

The crossguard should cover the head well.

The rear foot comes forward as you strike to become the lead foot, and this must point at your target. Otherwise, your strike will be weak and your balance poor as you hit. The rear foot must follow the strike in the sense that by stepping with it you generate extra momentum. Do not lean forward and allow yourself to become unbalanced.

"The zwerch *counters what comes from above."*

The *zwerchaw* breaks (counters) all downward strikes made from above. Do it like this: when he swings an *oberhaw* at your head, leap with your right foot close to his left side. In the leap, turn your sword so that the crossguard is high in front of your head and your thumb is downward, and strike at his left side using the short edge. Thus you will deflect his strike with your crossguard and simultaneously hit him in the head.

"Zwerch benymp was von tag her kümpt"

Merck, der zwerhaw bricht alle hew die von oben nyder gehawen werden und den haw tryb also wen er dir oben in hawet zu dem kopf so spring mit dem rechten fuß gen im uß dem hawe uff sin lincken sytten und im springen verwent din schwert mit dem gehultz houch vor deinem haupt das din doum unnden kome und schlach in mit der kurtzen schniden zu siner lincken sytten so vasch dü sinen haw in din gehultß und triffest in zu dem kopff

This is a counter to strikes from above, and it is by no means necessary to hit him in the head. Just striking his sword will give you excellent options for follow-ups. Step as soon as you sense his movement, and strike with strength. Remember to keep the crossguard high to guard your head.

Make sure that you step slightly to the side so that you come at his outside rather than straight on. By doing this you will strike the flat of his blade with your edge.

"Do the zwer *with the strong, note how the work is done"*

This is how you can work with strength from the zwerch. If you strike a *zwerch*, remember that you should strike with the strong of your sword against his sword blade. If he holds strongly against you, strike his sword with crossed arms, behind his blade, in the head, or cut him across the face.

"Zwer mit der storck den arbeit do mit morck"

Daß ist wie du mit der störck auß der zwer arbeiten solt und dem thun also / wen dü im mit der zwer zu hauest so gedenck daß dü im mit der zwer sterck deines schwertes starck in daß sin hauest helt den er starck wyder so schlach in am schwert mit gekreutzten armen hinder seines schwertß klingen uff den kopff oder schnyd in mit dem stuck dürch daß maul

After you have struck a zwerch *you will be in this position. From here you have the following options: you can make a direct drawing cut to his hands from this position.*

From the zwerch, *turn the sword by making a* duplieren *movement and in the turn strike him in the face. The difference here is that you must aim your point slightly up so as not to lose contact with his sword.*

You can also push your blade forward at his face with a forward cutting motion. Simply turn the sword so that you can lower your hands and cut forward with force in time with a small step forward.

Strike around at his blade or aim lower at his abdomen; again, a raising strike will be harder to displace. This variation can be done in time with a step back.

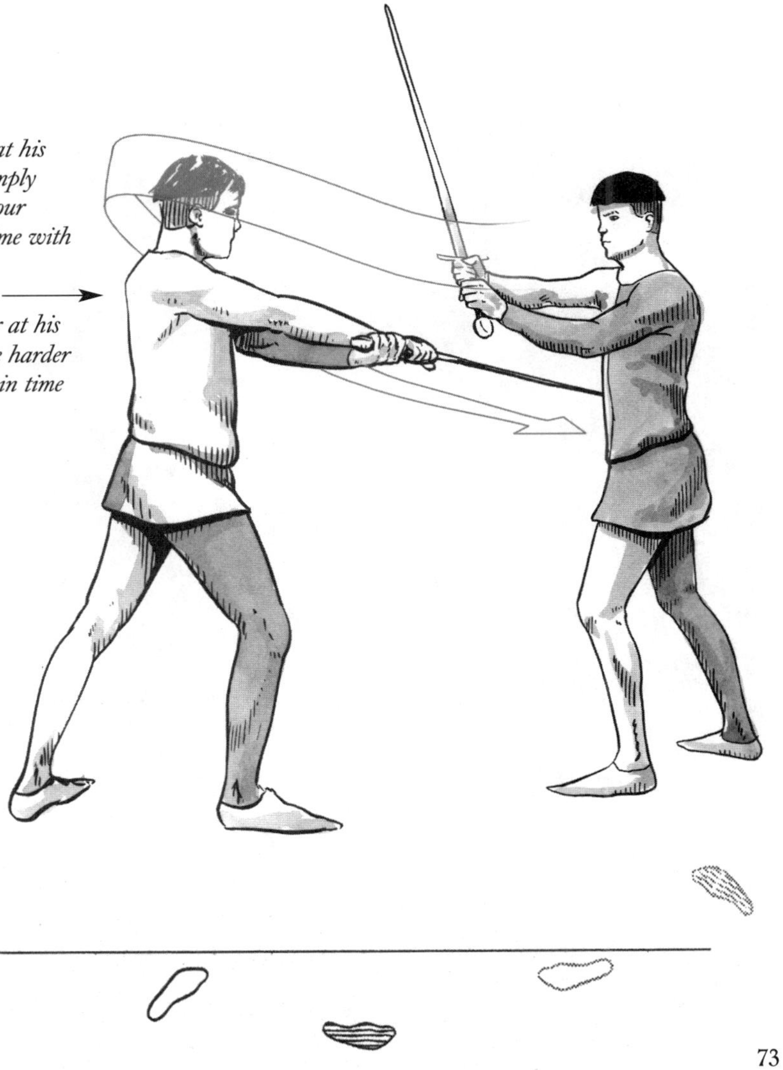

"Another technique from the zwerhaw.*"*

If you bind strongly with your sword against his from the *zwerch* and he holds strongly against, then thrust his sword downward on your right side using your crossguard. At once strike again a *zwerch* toward the right side of his head.

"Aber ain stuck uß dem zwerhaw"

Merck, wen du im uß der zwer mit der storck deines schwerts an sin schwert bindest, helt den er starck wyder so stoß mit deinem gehultz sin schwert von dir undersich uff dein rechten sytten und schlach bald mit der zwer wyderum gen siner rechten syten im zu dem kopffe.

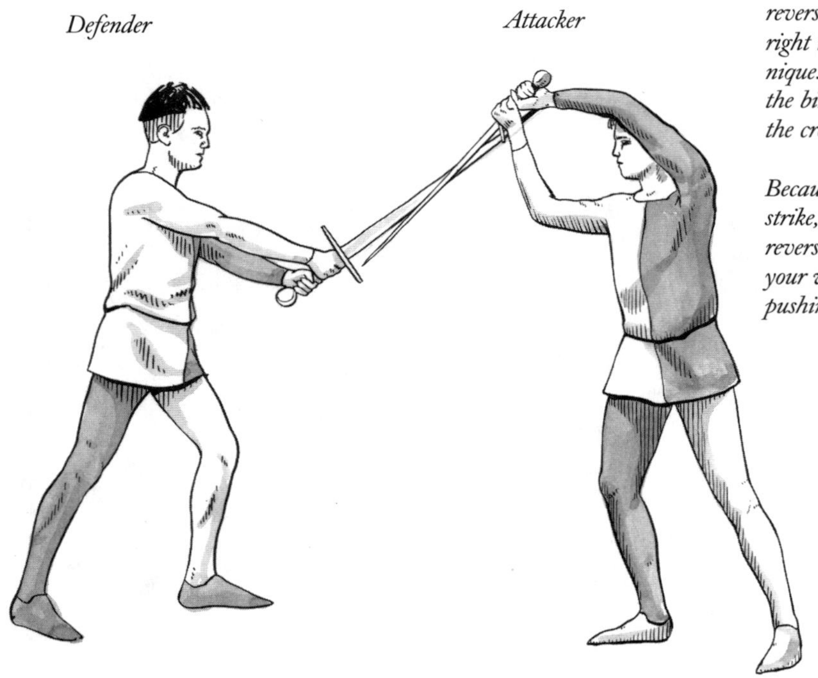

Note that the figures are reversed and the figure on the right is performing the technique. On the right: just after the bind as you begin to lower the cross.

Because he will resist your strike, you will find it easy to reverse and push him down. Use your waist, not your arms when pushing his blade down.

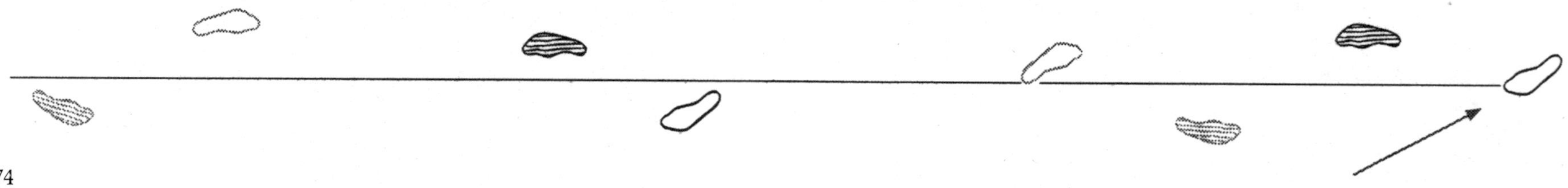

Once his sword is pushed down and to the side, take a step forward with the rear foot and a step back with the front foot. Thus you will create speed and momentum in the strike. Aim at his head or neck.

When you push his sword down, make sure that your point is aimed slightly forward and down, NOT excessively back and down.

Note that the distance is still the same; if you come too close you will not be able to strike a true blow. Maintain proper distance.

Take a step to add force to the strike, or strike while standing still–both work fine. Again, remember that your footwork must match reality. Reality will never match a set pattern of movement.

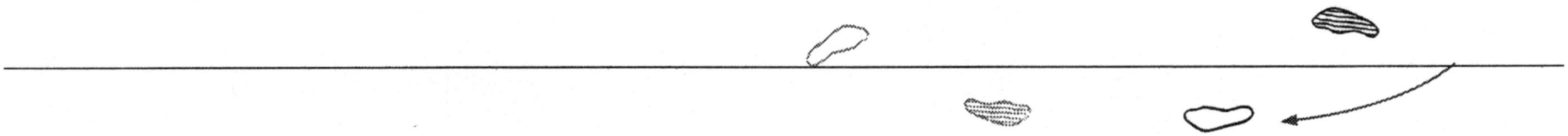

"Another technique with the zwerchaw."

If you bind your sword against his with the *zwerch* and he is weak at the sword, place the short edge on the right side of his neck. Then leap with your right foot behind his left and move in using the sword.

"Aber ain stuck uß dem zwerhaw"

Item wan dü im mit der zwer an sin schwert bindest ist da er waich am schwert so leg im die kurtzen schnyden zu seiner rechten sytten an den halß und spring mit dem rechten fuß hinder seinen lincken und rucke in mit dem schwert dariber

By lowering your crossguard and simultaneously moving it slightly to your right you can make it slip in behind his blade, i.e., between his sword and his body.

Take a large step forward or two shorter ones so that your leg is behind his. Destroy his balance by pushing forward with your body–not just your arms, they are too weak–before tripping him over your leg.

If you do not break his balance first, he can easily resist you and unbalance you instead. In all techniques that involve tripping, throwing, and the like, breaking the balance first is all important.

Push forward and to your right. At the same time you can push forward with your knee against his to unbalance him further or kick his front leg out from under him.

You can also use the sword as a lever to pull him to your right side in a circle over your leg. Pull your sword to your right side and down sharply. Think of how you move when paddling a canoe.

"Another technique."

If you bind your sword against his using the *zwerch* and he is weak at the sword, then push his sword down with the *zwerch* and place the short edge behind his arms on his throat.

"Ain ander stuck"

Item wen dü im mit der zwer an sin schwert bindest / ist er dan waich am schwert so truck mit der zwer sin schwert nyder und leg im die kurtzen schnyden hinder sinen armen vornen an den halß

This is basically the same entry as the previous technique. The difference is that here you employ a forward cut without trying to push him over. It is therefore considerably faster, but naturally of no use against an even lightly armored opponent.

You can also turn your blade like a hinge around his blade and cut his throat with a front horizontal cut as you leap forward.

When making the cut, step forward. Do not just extend your arms to cut. Use the body.

It can be done on both sides–that is, you can press his sword in the direction of the strike. Or change the direction and yank the sword to the other side with your crossguard. Thus you cut on the left and right side respectively.

Take a short step with the lead foot.

"A counter to the upper zwerchaw."

If you bind his sword with an *oberhaw* or another strike from your right side and he strikes a *zwerchaw* toward your other (left) side, counter with a *zwerchaw* to his neck under his sword.

"Ain bruch wider den obern zwerhaw"

Item wan dü im von deiner rechten sytten mit ainen oberen haw / oder sunst an sin schwert bindest schlecht er dan mit der zwer umb dir zu der anderen sytten so kom vor auch mit der zwerch under sin schwert im an den halß

This simple move takes you out of the way of his strike while at the same time you complete your own. Since he is striking a zwerchaw *you will not be able to hit his neck. But you will strike his arms at the level of his neck and catch your opponent in a position that is very hard to defend, especially against an incoming* zwerchaw.

If you also angle your sword slightly inward (away from you) you will be able to hit his hands or arms and you will be able to strike him from an even safer distance and position. When practicing this move take extra care to avoid a simultaneous hit. You must find the proper move and position that will let you hit him, but not the other way around.

As an alternative in stepping you can try to cross your left foot over your right as you strike. It is not a position to hold, but if you are fast and take a second step with the right foot, you can actually step almost behind him.

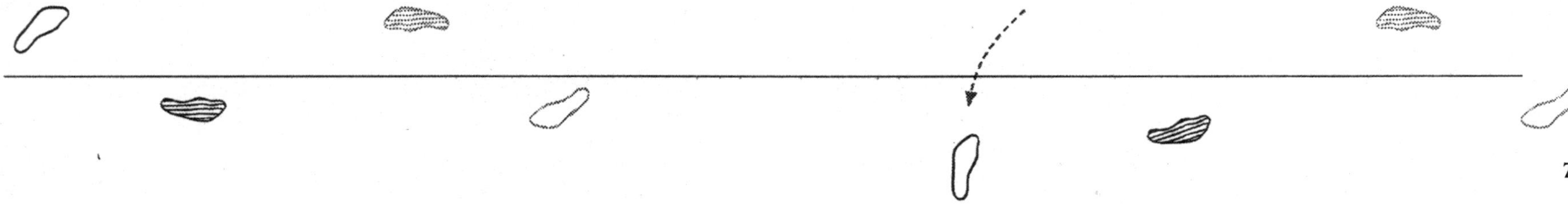

"Do the zwer *at the plough, to the ox strike hard."*

How one should strike at the four openings using the *zwer.*

This is how you can strike at the four openings with the *zwerchaw.* When you close with your opponent, leap toward him at the right moment and strike a *zwerch* at the lower opening on his left side. This is called to strike at the plough (*zum pflug schlagen*).

When you have struck a *zwerch* at his lower window, then at once strike a *zwerch* to the opposite side of his head. This is called to strike at the ox (*zum ochsen schlagen*). Then strike swiftly a *zwerch* at the plough and the ox, changing crosswise from one side to the other. Then you can free yourself from him with an *oberhaw* to the head.

"Zwer zu dem pflug/ zu dem ochsen hart gefuge"

Wie man zu den vier blossen mit der zwer schlachen soll

Daß ist wie dü in ainem zu gang mit der zwer zu den vier blossen schlagen solt das vernym also wan dü mit dem zufechten zu im kompst so merck wan es dir eben ist so spring zu im und schlag in mit der zwer zu der underen bloß siner lincken sytten das hayst zu dem pflug geschlagen

Wen dü im mit der zwer zu der underen bloß geschlagen hast so schlag bald uff mit der zwer im zu der andern sytten oben in zu dem kopff das haist zu dem ochsen geschlagen und schlach denn fürbaß behendlich ainen zwerch schlag zu der ochsen und den anderen zu dem pflug creutzwyß von ainer sytten zu der anderen und haw im do mit ainem oberen haw oben ein zu dem kopffe und zuch dich da mit ab

Striking a zwerchaw *at a lower window makes it necessary to lower the crossguard, otherwise you will only hit with the tip and do no damage. If your opponent defends with a* pflug, *close in and wrestle. If you try to strike him on the other side, you will get hit by a thrust. If he voids or deflects low, continue.*

If you have the opening, strike him in the head on the other side. Make a pass with your rear foot and let your "new" rear foot move slightly as you strike to generate more momentum.

If your opponent parries both strikes, try to hit him with a vertical schaytelhaw. *The human mind is easy to lure into pattern thinking–if we are attacked by two similar attacks, the mind has a tendency to anticipate a third. Therefore it is relatively easy to hit with a different strike as a third one.*

"When you leap with the zwerch, *protect the head."*

When you strike with a *zwerch*, always leap toward his flank, or to be precise, to the side that you wish to strike. Then you can hit him in the head. Remember to cover your head with your crossguard while leaping.

"Waß sich wol zwerch mit springen dem haupt geferet"

Das ist das dü mit ainem yden zwer schlage wol uß solt springen im uff die sytten / do dü im zu schlagen wylt so mag stü in wol treffen zu sienem haupt und wart daß dü in dem sprung oben vor dinem haupt mit dinem gehultzen vol bedeckt syest

Correctly executed zwerchaw.

For this technique to work it is important that you remember to cover your head with your crossguard.

This is too low, and allows your opponent to tip his sword over your crossguard and hit you.

You can of course do a zwerch *on either side. If you are right-handed, however, you will not be able to strike with a* zwerch *with the short edge on your left side without shifting your grip. This is not good, so on the left side you strike with the long edge instead (at least if you are crossing over from an earlier strike on his left).*
If you initiate a zwerch *on his right, you can strike with the short edge. Try both variations.*

Incorrect zwerchaw*; the crossguard is too low.*

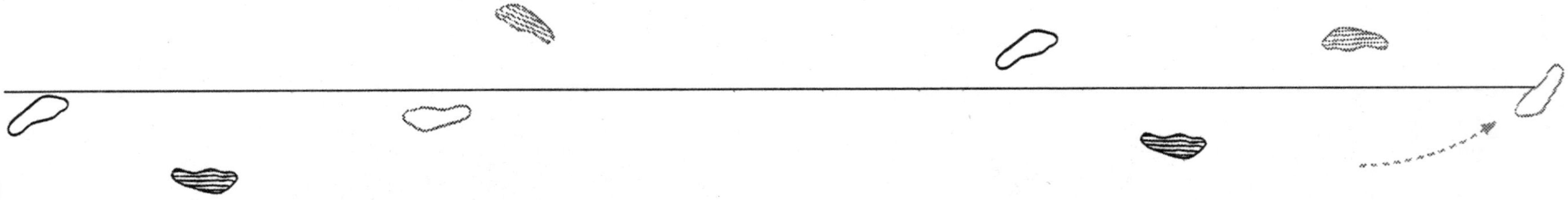

Feler and Verkerer

The feler and *verkerer* belong to the same family as the *duplieren* and *mutieren* movements: they are specific movements that can be employed in specific situations. In that sense they are, at least as they are shown, not principles. On the other hand, these movements are examples of different kinds of feints, and feints as such are very much a principle. So let us say that these are Liechtenauer's and Ringeck's illustrations of an idea rather than illustrations of exactly how to execute them.

A feint is dangerous by its nature both to the man who executes it and to the receiver. It is easy to practice feints in class and then tend to rely on them too much in free play. This is wrong; the simple principles of fencing always come first, and the feint should be used with caution and restraint.

The feler, or feint proper, is a simple move that can be made in many variations. The basic idea is simply to pretend to strike at A and then change the strike and hit B instead. The difficult part with a feint is that it will expose you dangerously; if your opponent sees through the feint he will have a golden opportunity to hit you. And since you are in the process of changing the strike, you will be very hard-pressed to counter his attack. Secondly, you must try to make the feint as small as possible because a larger movement will be harder to change and will take longer to change, thus giving your opponent more time to counter or void. At the same time, the feint must be clear enough to capture your opponent's interest. A delicate balance.

The verkerer is a different kind of feint. It begins with a fake thrust at the opponent's face. As he parries, he will expose one of his arms (which one depends on how he parries). His arm or elbow can then be grabbed and he can be thrown or pushed, or you can disarm him. The important thing to note is that the *verkerer* means to turn your sword to thrust and then use the opening created. This can be done from several positions. You do not have to push his arm to the side; the *verkerer* is a way to create an opening from a fake thrust, or, for that matter, a thrust that was deflected. You are parried and with the *verkerer* you regain the *vor* and the initiative.

"He who does the feint will hit below as he wishes."

All fencers who rely on deflecting or parrying (*versetzen*) will be deceived and defeated with the feint (*feler*). When you close with him, then feint any *oberhaw* at his left side. When he tries to displace this you can hit one of the lower openings with ease.

"Feler wer wal furet und unden nach wunsch ruret"

Daß ist mit dem feler werden alle fechter die da gern fersetzen ver fyrt und geschlagen daß stuck trib also Wan du mit dem zu fechten zu im kompst so thu alß ob du in mit ainem fryen ober haw zu siner linck sytten so ist er unnden nach wonsch geruret und geschlagen

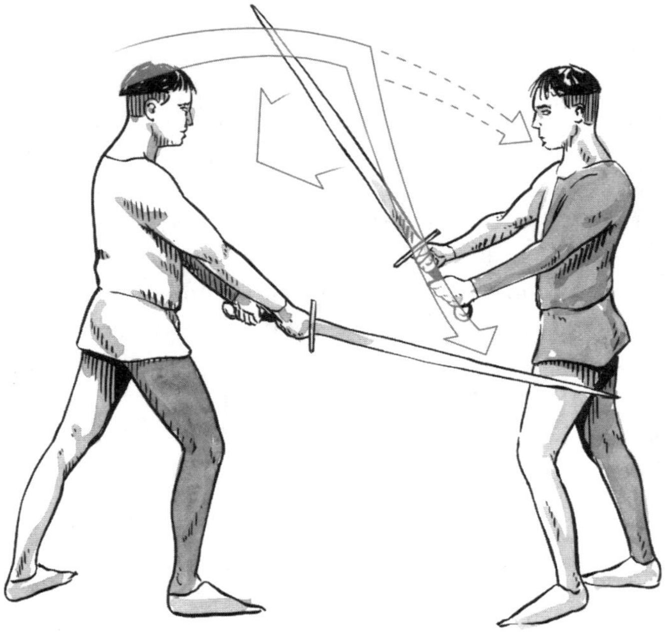

The feint can be more or less obvious, just be sure to aim high. You can feint without moving, or by taking a step with the left foot, leaving the right one behind to help with the real strike.

Pull the sword back and take a step as soon as you have your opponent committed to the attack. Swing the sword around and take a step with the right foot forward and slightly to your right.

It is easier to strike high in the lower window than very low, and you do not need to expose yourself so much if you aim higher in the window.

You can, of course, do this on either side. If you want to strike around (high right/low left) allow the swords to make contact before you change. Otherwise you must strike loosely or else hurt your shoulders when changing. Do, however, let the first strike be weak and with little commitment; that makes it easier to change your direction and strike around.

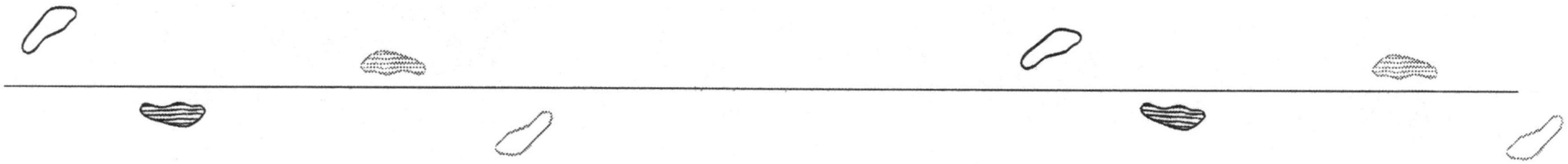

"The verkerer *forces the running through with the wrestling, take the elbow and jump in balance"*

When you bind someone at his sword with an *oberhaw* or *unterhaw*, turn your sword so that the thumb is down and thrust him in the face from above, thus forcing him to parry the thrust. Then grab his right elbow using your left hand and place your left leg in front of his right and push him over it. Or move in under his weapon and start to wrestle (*ringen*) as will be described in detail later.

"Verkerer zwinget durch lasser auch mit ringet Den elenbogen gewisse nym spring im in die wage"

Merck daß stuck soltü also tryben wen dü in mit ainem under oder oben haw an sin schwert bindest so verker dein schwert das din doume unde kome und stich in oben in zu dem gesichte so zwings tu in das er dir versetzen muß und in der versetzung begriff mit der lincken hand sin rechten elebogen und spring mit dem lincken fusse für sinen rechten und stoß in also dariber oder lauff in mit dem verkerer durch und ringe alß dü in dem durch lauffen her nach wirst finden

Verkerer *translates as backward, upside down, turn, or turning.*

Just slide the sword up along his blade as if you want to place a thrust in his face. You may even make a thrust if he does not deflect it.

As soon as he parries this thrust, grab his elbow. If he uses a kron *to displace, do the technique on the other side using your right hand and leg. As you grab him, take a step forward and turn at the waist to push him to the side. Do not use your arms; they are too weak. The movement should be fluid and explosive.*

When you leap in you must control his elbow as you push his sword to the side. This prevents him from swinging back to strike or seize you.

Once he is turned to the side you are free to dispose of him. Be sure to do it in such a way that you maintain control of his elbow.

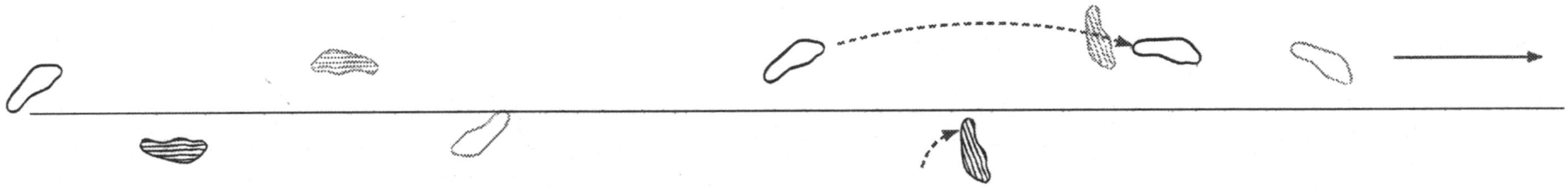

"Feint twofold; if you hit then make the cut."

Note that it is called the twofold feint, since you must change twice while closing in on him. When you come in close, leap with your foot toward him and fake a *zwerchaw* to his left side at his head. Then change the direction of the strike and hit the right side of his head.

"Feler zwyfach trifft man den schnit mit macht"

Merck das haysst der zwyfach feler darumbe das man in ainem zu fechten zwayerlay ferfürung daruß tryben sol die ersten tryb also wen dü mit dem zu fechten zu im kompst so spring mit dem fuß gen im und thun alß dü im mit ainem zwer schlagen zu siner lincken sytten zu dem kopff schlagen welest und ferzuck den schlag im zu siner rechten sytten an den kopff

If your feint is too committed and strong, you will not be able to strike around quickly.

As you strike around, keep the circle of the sword small. Keep in mind that the less the sword point moves, the faster the change will be, of whatever kind. So pull the sword up and around to avoid creating too large a circle.

You can step with either foot. Since the real strike is to be on the second side, try stepping to place yourself in a good position for that, i.e., with the left foot forward. Otherwise it works well with stepping with the right foot also. Redirect your strike to the side you are stepping. The hard part is making the fake look believable to your opponent.

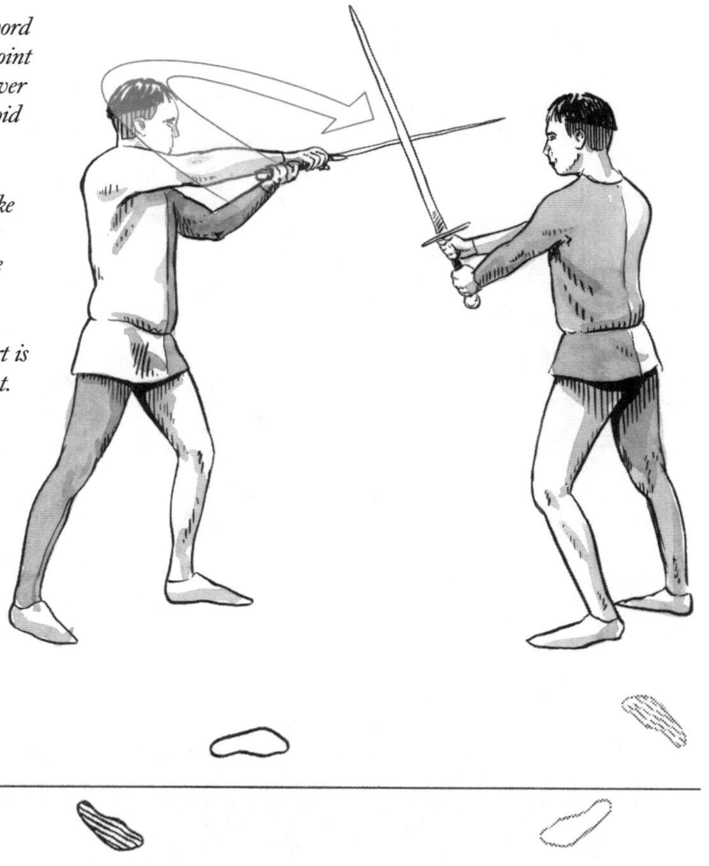

"Twice further on, step to the left and do not forget the cut."

As described above, when you strike him on the right side of the head from the first feint, strike him again at once in the same place. Then move the short edge, without crossing your arms, over his sword and leap toward your left side and cut him with the long edge across his face.

"Zwyfach es fyrbas schryt in linck und biß nit laß"

Das ist wan dü im mit der ersten verfyrunge zu siner rechten sytten zu dem kopff geschlagen haust / alß am nesten gemelt ist So schlach bald wyderumb im zu der rechten sytten zu dem kopff und far mit der kurtzen schnyden mit auß gecreutzten armen über sin schwert und spring im lincke das ist auff dem lincken sytten und schnyd in mit der langen schnyden durch das mal

A zwerch *to the left side, leading with the other leg.*

Strike around to his right side.

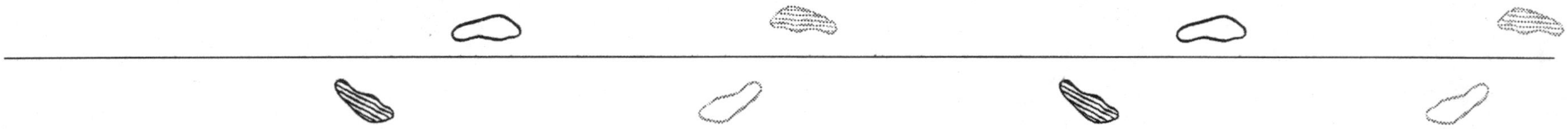

Pull back to charge with a second strike in the same place, making sure that you keep your balance as you land the second strike.

After the strike, turn the sword somewhat so that the long edge is aimed at his face, like doing the duplieren *backward. Push the long edge at his face as you leap forward and cut with force while leaping.*
Step with the left leg or make a cross-step with your right foot.
This movement should be well forward and to the side.

Schiller

The schiller, or squinting strike, is one of the *meisterhaw* and also one of the deflections. Ringeck points out that it is supposed to be used against the *pflug,* or against aggressive and unskilled fencers who rely mainly on brute-force attacks. The major difference in how we have interpreted it here is that it is done to try to keep the crossguard low and the point up, and to use the body rather than the hands alone to generate force, which is what you are forced to do if you hold the crossguard high. The strike can still be delivered quite high, and it is possible to hit your opponent's shoulder with it. That said, it can be done with the hands high and the point low, and there are plausible reasons for doing the technique in that way.

If we look at later manuals (by Meyer, for example) we see a picture that look like a *zornhaw* struck with the short edge and from the side (almost a *zwerch*). Unfortunately, Ringeck's text does not clarify the issues. The major problem in interpreting this technique is that to do it in Meyer's way is rather pointless; you can get better results by using a *zornhaw* or a *zwerchaw.* It seems an unnecessarily complicated solution. So we have suggested here another interpretation of the strike that satisfies the demands of the text, but as a unique and rather useful movement. No doubt there is much more to be said on this matter.

It is very important that you strike his sword FIRST and at the weak part of his sword using your short edge. You can step in several ways as you do this, and after striking his blade you can strike or cut his body. Against the pflug, *knock his blade down and to the side as you step.*
As a follow-up, thrust forward or make a drawing cut to his face or throat.

"The schiller *breaks what the ox strikes or thrusts; if someone changes through,* schiller *will rob him of it."*

Note that the *schiller* is a strike that breaks the strike or thrust of the clumsy or unskilled fencer who relies only on strength. Do it like this: When he strikes at you from his right side, strike from your right side using the false edge against his strike–with your arms extended against the weak of his sword–and strike him on his right shoulder. If he changes through, thrust him in the breast with extended arms. You can also strike like this if he is in the position of the plough, or if he tries to thrust at you from below.

"Schiller ein bricht waß bufler schlecht oder stycht wer wechsel trawet schiller in dar uß beraubet"

Hie merck der schiller ist ein haw der den buffeln die sich maysterschafft an nemen mit gewalt in bricht in hawen und in stechen und den haw tryb also wan er dir eben ein hawet von siner rechten sytten so haw och von dener rechten sytten mit der kurtzen schnyden mit uff gerechten armen ge sinen hawe in die schwech sines schwerts und schlag in uff sinen rechten achsel wechselt er durch so schyß in mit dem hawe lang in zu der brust und also haw ach wan er gen dir stat in der hutte de pflugs oder wen er dir unden zu wyll stechen

Step forward and out toward the side you wish to strike. You must adapt your step to fit the distance and direction of your opponent and his moves. Instead of striking the shoulder you can use this strike to hit his hands very efficiently, and follow with an upward thrust to his chest or face.

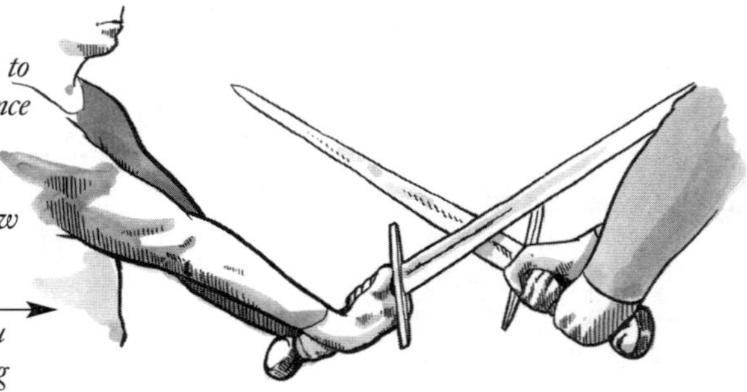

Against a thrust from below, as a follow-up you can cut or try to knock your opponent over using your sword as a lever while tripping one of his legs. Ideally you should hit his weapon and body simultaneously with the same strike. This is very hard to do.

As an alternative that complies with the text, strike his blade at the weak and then duplieren *to his shoulder or head. Notice that you must twist your wrist considerably. This is so your sword will automatically place itself in roughly a 45-degree angle, protect your body from your opponent's sword, and enable you to hit him at the same time.*

"Make the schill *short, change through and thrust at him."*

Note: You must at all times keep your opponent in your field of vision. If he tries to strike short, you will see this if he does not extend his arms in the strike. Then you should strike at once and guide the point of your sword under and past his sword and stab him in the face (or breast).

"Schill kurtß er dich an durchwechsel er sigt im an"

Merck das ist ain lere das schillern solt mit dem gesichte und gar eben sechen ober kurtz gen dir vicht das solt by dem erkenen wan er dir zu hawet und sin arm dem haw nicht lanck streckt so haw och und far in dem haw mit dem ort under seinem schwert durch und stiche in zu dem gesicht

Because he will not extend his arms, it is relatively easy to change through and put the point of your sword below his arms. You do not really strike, but very quickly use the schiller *to get the sword into an advantageous position.*

Step quickly forward with your lead foot and thrust using your body for force, extending your arms only for reach.

"Make the schill *to the point and take the throat without fear."*

Note: The *schiller* breaks the *langen ort* thus: When he stands in front of you and aims the point of his sword at your face or breast with extended arms, then put your left foot forward and look at the tip of his sword. Feint a strike at the point of his sword, then strike powerfully on his sword using the false edge. At the same time take a step forward and to your right, thrusting the point of your sword at his throat.

"Schill zu dem ort und ny den halß on forcht"

Merck der schiller bricht den langen ort und den tryb also wen er ge dir stat und helt dir den ort usß gerachten armen gen dem gesychte oder der brust so stand mit dem lincken fuß fur und schill mit dem gesicht zu dem ort und thun alß dü im zu dem ort hawen welest und haw starck mit der kurtzen schnyden uff sin schwert und schuß im den ort / darmit lang in zu dem halß mit ainem zutrytt des rechten fuß

Immediately after you have beaten his blade aside, move the pommel (as in duplieren*) in under your right arm, and in one movement strike at his face or neck. Or you can aim the point at his face or throat and step forward with your lead foot. The strike will be faster.*

"Make the schill *to the head above; you will hurt the hands."*

If he strikes with an *oberhaw*, watch his head as if you intend to strike him there. Then strike with the short edge against his strike and at once strike with the tip of your sword along his blade at his hands.

"Schill zu dem oberen haupt hende wilt dü bedebren"

Merck wen er dir oben will in hawen so schill mit dem gesicht alß dü in uff das haupt wylt schlagen und haw mit der kurtzen schnyden gen sinen haw und schlag in an siner schwertz klingen mit dem ort uff die hend

As you strike, keep in mind that by pulling the sword up at the end of the blow you end up in the ochs *and can finish with a thrust. As you strike you can make a pass with your rear foot. This will also set you up nicely for a thrust and at the same time allow you to close the distance between you and your opponent, thus rendering any strikes he may make less dangerous.*

As your opponent strikes, you strike a schielhaw *at his sword. You are hitting it from the side, so little force is needed to turn it to the side and miss. If you commit too much, you are in danger of overextending yourself.*

To hit his hands, simply duplieren *and strike with the tip of your sword. From there you can also step in and push his hands down or wrestle him.*

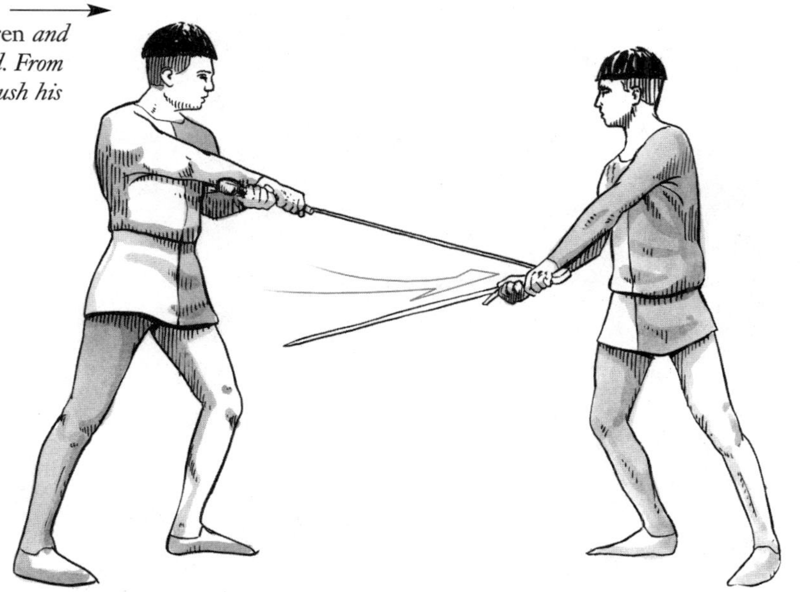

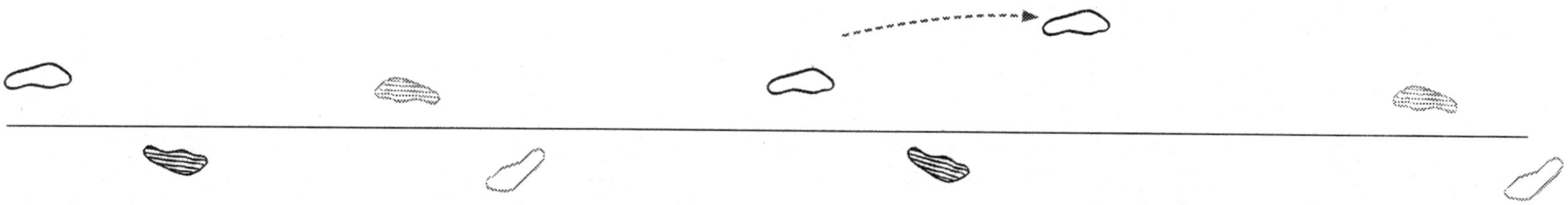

Schaytler

The schaytler is another one of the *meisterhaw*, and together with the *zornhaw* the most important strike. It is simple and can be executed with great speed. The strike, as such, is vertical and can be done either as a long full strike that reaches from your opponent's head to his groin or as a shorter strike that is aimed primarily at his head.

This second strike is done with the hands held high, as we will see later, while the full strike brings the hands down in a curve toward the ground. The *schaytler* also breaks the guard of the fool (*alber*) by attacking as high as possible without having to close into the distance of the *alber*'s threat from below. The *schaytler* is also very efficient as a counter aimed at the hands or head of an opponent in the *vom tag* position when the sword is held at the shoulder.

Schaytel can mean scalp or parting (as in splitting something). So the scalp-parting strike might be a good translation. It is so named because it is done vertically and often aimed at the head.

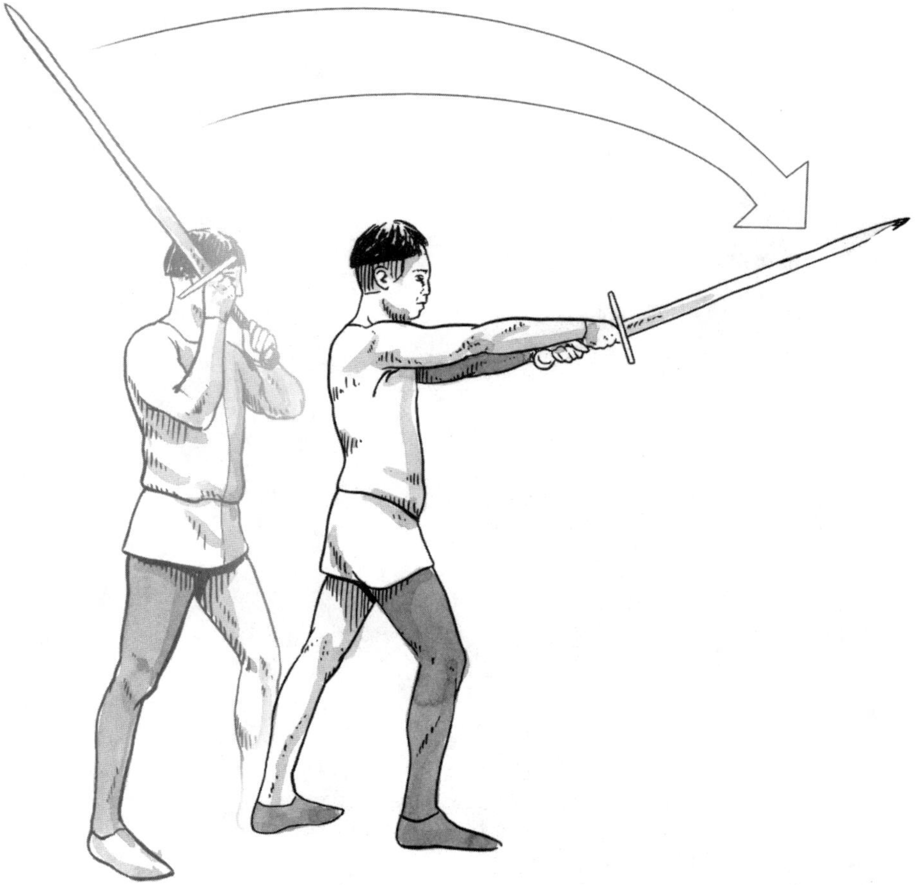

"The schaytler *is a danger to the face"*

Note: The *schaytler* is aimed at the face or breast. Do it like this. When he is in the *alber* position, strike him with the long edge from above to below. Keep your arms high and keep the point at his face.

"Der schaytler dem antlytz ist gefer"

Hie merck der schayteler ist den anlytz und der brust gefärlich den tryb also wen er gen dir stat in der hut alber so haw mit der langen schnyde von der lange schayttlen oben nyder und belyb mit dem haw hoch mit den armen und heng im mit dem ort ein zu dem gesychte

For the schaytler *to work your arms must remain high in the strike. This ensures that you do not pull the blade in a descending arc, a movement that would make you hit later than if your arms remained high. Also stretch forward with your shoulders and keep the body at the same distance as when in the ready position* vom tag. *Step in any convenient way to hit him. You can move forward, remain still, or even take a step back.*

"With its turn, very dangerous to the breast."

If you aim the tip of the sword at his face using the schaytler and your opponent pushes your sword point over him by lifting his crossguard above him, then turn your sword and lift your crossguard high over your head and put the point below on his breast.

"Mit siner ker der brust fast gefer"

Das ist wen dü im den ort mit dem schaitler oben ein hengst zu dem gesicht stost er dir denn den ort in der versatzung mit dem gehultz vascht yber sich So verker dein schwert mit dem gehultz hoch fyr din haupt und setz im den ort unden an die brust

The kron *(or crown) can possibly be done in two ways. Below is with the point up, and to the left with the point to the side. I have interpreted the technique in such a way that after you have struck a* schaytelhaw*, your opponent uses the* kron *to avoid the point of your sword. A second reason for interpreting the technique as being applied to the side is that no matter what you do, you cannot place the point of your sword on his chest by lifting your crossguard upward. The point will still be caught on his crossguard. The only way that seems to work is if he slants his sword to the side with the crossguard over his head. Possibly the technique is intended to include both variations! Lift high to block or parry, and push to the side to counter because it is just as impossible to attack if you hold your sword straight up.*

Alternative kron

The thrust can be made as in the picture, or by pulling your sword back as you go high so that you clear his blade, and then all at once thrust at his chest with a step forward with your lead foot.

"What comes from him, the kron *will catch."*

If you strike a *schaytler* and he deflects it with the crossguard high over his head, that displacement is called the *kron*. From this you can also move in under his sword.

"Waß von im komp / die kron das abnympt"

Merck wan dü im mit dem schaittler oben ein hawest / versetzt er mit dem gehultze hoch ab ob sinem haupt die versatzung hayst die kron und laufft dir do mit eyn

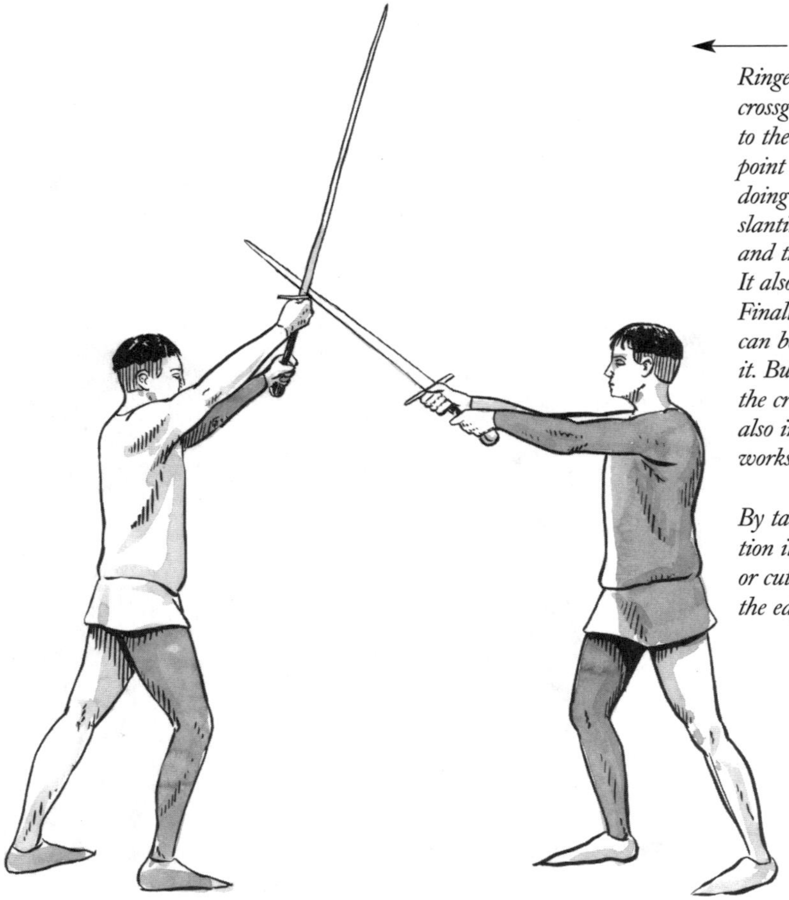

Ringeck does not specify how to lift the crossguard over your head. You can do it to the side or lift the sword straight up, point upward. The advantage we see in doing it to the side is that the forward-slanting sword is seen in many manuals, and that this is a plausible interpretation. It also allows for several follow-ups. Finally, by doing it like this the technique can be done exactly as Ringeck describes it. But note the variation to the left with the crossguard lifted straight up–this is also in agreement with the text, and it works with the techniques.

By taking a step forward in either position it is possible to come close to grapple, or cut at his arms or push him away with the edge.

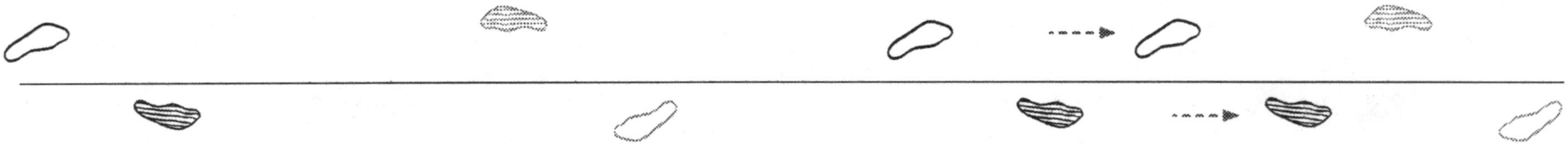

"Cut through the kron*; you will break through hard, push away the techniques, and take them off with the cut."*

If he breaks your *schaytler* or another kind of *oberhaw* with the *kron* and tries to move in, cut under his arms and push up to break the *kron.* Turn your sword from a low cut to a high one, then you can free yourself from him.

"Schnid durch die krone / so brichst du sy hart schon die stuck drucke mit schnitten sy ab zucke"

Merck wen er dir den schaittler oder sunst ainen haw versezt mit der kron / und dir da mit ein laufft / so nym den schnit under sin henden in sin arm und truck vast uber sich so ist die kron wyder gebrochen und wende din schwert uß dem underen schnit in den oberen und zuch dich da mit abe

As you sense that he is about to move in, push your crossguard toward his body: that will place the edge on his arm. Cut by pushing your sword upward. You can also cut him by placing your sword more horizontally and making a drawing cut to the side.

Regardless of whether his sword is up or to the side, step in and push your crossguard forward toward his arms. This will let you reach him without losing contact with his sword and opening yourself up to a counter. Pull your sword down in the cut.

Take a step with your lead foot and cut upward and forward. This will push him back and cut his lower arms as well.

If his point is upward, place the long edge on his arms from below, push the cut up and turn the sword over his arms to cut him from above. To free the sword, push the pommel forward, which will move the point back. Then push the pommel to the side and cut.

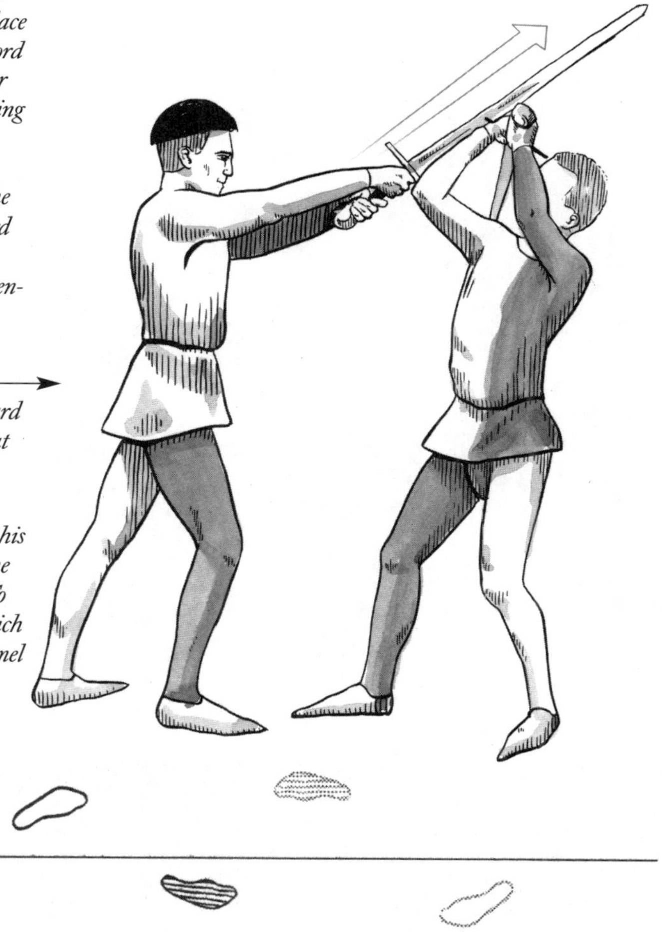

Vier Versetzen

Versetzen means "deflection" and is Ringeck's central tactic for dealing with attacks. At the same time, you should as much as possible abstain from relying on only deflecting attacks, and instead try to make well-timed counters while moving to avoid the attacks. (It is important to understand "deflect" [*versetz*] as meaning an interception and not a static blocking action.) This is sound reasoning because if you make a parry or deflection, both you and your opponent will make the same action–but his can hurt you while yours only negates his threat. You do not threaten him. In a sense you "lose tempo," as they say in chess. This is also connected to the concepts of *vor* and *nach*; by simply waiting to deflect at all times you will remain in the after (or *nach*) and lack initiative.

Versetzen is a powerful tool for dealing with attacks or opening up positions, but it will not win a fight. It is also worth noticing that Ringeck seems to primarily intend the *versatzungen* to be used against the guards or very specific attacks. A guard is a position where you are well placed for both attack and defense and you are difficult to reach, so the *versatzungen* can also be seen as the way of safely taking the *vor* from your opponent if he is waiting in a guard. In the last paragraph Ringeck states that you should fence like your opponent: this of course is also sound advice. But consider that it means that you should not try a thrust against a strike in hopes of deflecting it, not that you must never change to other tactics. It is a caution. All of this is linked to the concept of *nachrayßen*, where you are told to strike if he strikes, and thrust if he thrusts. You follow your opponent if he outtimes you–this is how you regain the *vor* from the *nach.*

"There are four deflections that are very hurtful to the four guards; beware of deflections because they will trouble you greatly."

You have been told before that you should only fight from four positions. Now you will learn the four deflections, which are four strikes.

THE FIRST STRIKE is the *krumphaw*: this breaks the guard of the *ochs.*

THE SECOND IS the *zwerhaw*: this breaks the guard of the *hut / vomtag.*

THE THIRD IS the *schielhaw*: this breaks the guard of the *pflug.*

THE FOURTH IS the *schaytler*: this breaks the guard of the *alber.*

Beware of all displacements used by bad fencers. Note: When he strikes, you should strike too, when he thrusts, you should thrust too. How you will strike and thrust you will find described in the five strikes and the displacements.

"Vier sind versetzen die die leger auch ser letzen vor versetzen hiet dich / gesicht es ser es myt dich"

Merck dü hast vor gehort das dü alain uß viern legern vechten solt So solt dü och nun wyssen die vier versetzung das sind vier hew

DER ERST HAW ist der krumphaw der bricht die hut des ochsen

DER ANDER IST der zwerhaw der bricht die hutt vom tage

DER DRITT IST der schyllhaw der bricht die hut deß pflugs

DER VIERDT IST der schaittler der bricht die hut alber.

Und hyt dich vor allen versetzen die die schlechten vechter tryben und merck wen er hawt/so haw och/und wen er sticht so stych och /und wie dü hawen und stechen solt/das findest dü in den fünff hewen und in den absetzen geschryben

KRUMPHAW AGAINST OCHS

To break the ochs, *step out to the right side and aim a* krumphaw *at his blade. Ringeck does not say if you are to strike the sword or the man. Both work well, but the sword is closer and following up is an easy matter after the threatening point is removed. Of course, the* krumphaw *may be executed with either the long or the short edge.*

Do a follow-up at once, stepping to your opponent's blind side away from his sword, and strike at his face, head, or legs. Be careful that he does not wind his sword to thrust at you from below or the opposite side.

It is also possible that Ringeck's meaning is to execute a simultaneous deflection and counterstrike in one continuous action. The technique can be done regardless of which side he is on.

ZWERCHAW AGAINST VOM TAG

To break vom tag, *simply step forward and to the side, and aim a* zwerch *at his left arm, head, or shoulder. Should he strike against your strike, hold strongly and work from the bind.*

You can also target his hands–your strike will then be farther out. The crucial moment is the step, which must be fast and decisive.

If he is holding the sword at his shoulder, aim for the left side of his face or his left shoulder.

SCHIELHAW AGAINST THE PFLUG

To break the pflug, *you can strike the* schielhaw *in several ways, depending on what you wish to do next. The difference lies in how far up the blade you strike–in other words, how close you will come to your opponent.*

(An easy follow-up is to simply take a small step forward and thrust at his face, or make a drawing cut to his neck on the right side.)

SCHAYTLER AGAINST ALBER

To break the alber, *strike with your hands and elbows kept quite high. But be careful not to snap your elbows upward: you can hurt yourself quite badly that way. Hit with the tip of your sword to retain a maximum distance between you and your opponent. You can also do this if he strikes from below from* alber. *The principle is that your upper strike has a shorter distance to travel and will hit before his strike from below (provided you do not lower your hands too much in the strike, since this will take more time and put them in danger of being hit by a rising strike).*

Either step forward to strike or, if he is within reach, simply strike. Be aware that your lower body may be attacked with a rising strike. You can hit with your counter-strike to the head, or lower your hands in the strike and counter to his hands or arms.

"If you are deflected above, when that happens hear what I advise. Move quickly away and strike with surprise."

If your strike is deflected, then note: if your *oberhaw* is deflected, remain in the deflection and move the pommel over his forehand and rip it down so that you simultaneously strike him on the head with the sword.

"Ob dir verstzt ist und wie das dar komen ist höre was ich rate rays abe haw schnel mit dratte"

Das ist wie es dar zu komen ist das dir versetzt ist worden so mercke versetzt man dir ainen ober haw so var im in der versatzung mit dem knopff vir sin vorgesetze band und rayse da mit under sich und mit dem rayse schlage in mit dem schwert uff den kopff

Against an oberhaw *you will most likely cross swords with your pommel to the right (if you are right-handed).*

The idea is to lift your sword, thus lifting his. Move your pommel slightly to your right as you lift your sword. Let your blade turn around his blade, making sure that you maintain contact. This will ensure that you retain control of your opponent's movements.

This will generate the space you need to pull his sword down. It works on both sides and is best with a pass forward to his side.

This footwork leads to the footwork on the following page.

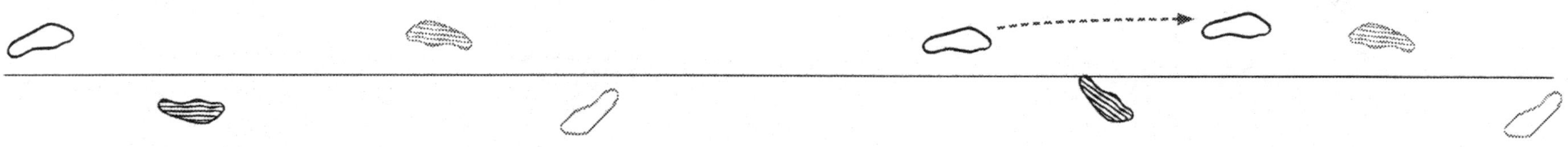

Note that if your grip is too short, your arm will push his arms away as you try to capture him. Move your rear hand closer to the crossguard (or use a sword with a longer grip).

The hooking motion with the pommel is done behind his front hand. You must slam the pommel into his hand and fingers; you do not do this gently. In the same movement you begin the strike, so it should be a single movement, not two. The strike is a curve, pulling the blade down and toward you.

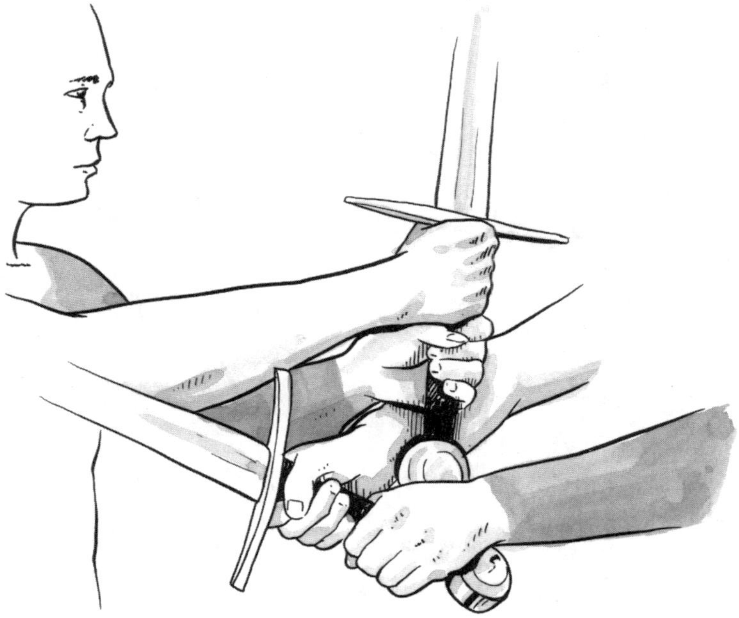

"Another technique from the displacement."

If you strike with an *unterhaw* from the right side and he hits your sword with his own, pushing your sword down so that you cannot get it up, then quickly move your pommel over his sword and strike him with a snapping movement in the head using the long edge. Or if he hits your sword on the left side, strike him with the short edge.

"Aber ain stück wider die versätzung"

Merck wen dü hawest ainen under haw von der rechten sytten fellet er dir dan da mit dem schwert uff das dine das dü do mit nicht uff komen magst so far behendlich mit dem knopff uber sin schwert und schlag in mit dem schnappen mit der langen schnyden zu dem kopffe oder velt er dir gen diner lincken sytten uff das schwert so schlach in mit der kurtzen schnyden

Your opponent steps in and stifles your strike with a strike of his own, pushing your sword down.

As your opponent presses your sword down, lift the pommel and move the sword in a backward arc. There should be a feeling of lifting the sword upward while swinging it around. Try to keep your shoulders and upper arms as immobile as possible and work with your hands and lower arms.

You can step in several ways or still continue in the strike. By stepping to the open side you will generate distance, but the angle will be less favorable for your own strike. However, you lessen the chance of your opponent's sword hitting you with a rising strike, for example.

Swing around in an arc.

Complete the technique by striking him in the head.

"Set upon the four ends, stay upon them, learn to end it."

If you strike an *oberhaw* from the right and wish to end the fight quickly, then note: when he deflects you, strike at once around to the other side with a *zwerchaw*. Grab your sword blade in the middle of the blade with your left hand and put the point in his face. Or attack one of the other openings, whichever you find easiest to get at.

"Setz an vier enden / blieb dar uff lere wiltu enden"

Wen dü im von dener rechten achseln oben hawest / wilt dü denn bald mit im enden so morck wen er versetzt so schlach bald umb mit der zwer und begriff mit dem schwert mit der lincken hand miten in der klingen und setz im den ort in das gesycht oder setz im an / zu den vier blössen zu welcher dü am basten komen kanst

When you hit him with the second strike you will not be able to do a zwerch *with the original short edge. In this case, you will actually make a horizontal strike with the long edge, but with your hands crossed. Step forward with the left foot and strike with the* zwerch *to the side.*

When you take your sword to thrust, grab it in the middle of the blade, and in the thrust use only your left hand to guide it. The rear hand must supply the strength. Do not make a pass as you thrust, but step forward with your lead foot: this saves time and makes the thrust more accurate. Have a feeling of falling forward with the body instead of stepping and thrusting.

"Another technique from the deflection."

When you have thrust at his face with the halfsword and he deflects this, strike him with the pommel on the other side of his head. Or leap with the right foot behind his left and move the pommel over his right shoulder, around his neck, and pull him over your right leg.

"Aber ain stuck wider die versatzung"

Item wan du im mit dem halben schwert den ort in sin gesicht setzest / versetzt er dir das so stoß in mit dem knopff zu der anderen sytten zu seinem haupt Oder spring mit der rechten fusß hinder sin lincken und far im mit dem knopff uber sin rechten achseln vornen umb den halß und ruck in da mit uber din rechte bain

Continuing from the previous technique, if he parries your thrust, note that he can lead with either leg and so can you.

Step around him and strike him in the head. Since he is deflecting your thrust, he will turn his body to the side, making it easier to hit him.

Push his sword to the side using your own as you step, and use the force of your body (not just your arms) to move him to the side. In the same motion, strike him in the head with your pommel. Try to make it one continuous movement. That way you will generate enough force to crush his skull.

It is not essential that you step behind his leg to pull him over; the momentum of the turn is terrific if done with force and intent. Adapt to the situation and remember the principle of the movement: to use the sword as a lever and your leg as the fulcrum.

A detail of the hooking motion. You must have a fairly wide grip to do this. After placing the grip, turn at the waist and use your body (not just your arms). But pull your arms toward you with force as you turn to the side. Try to pull your opponent sideways and down, not just to the side. If done properly you can send him flying. Practice this with caution.

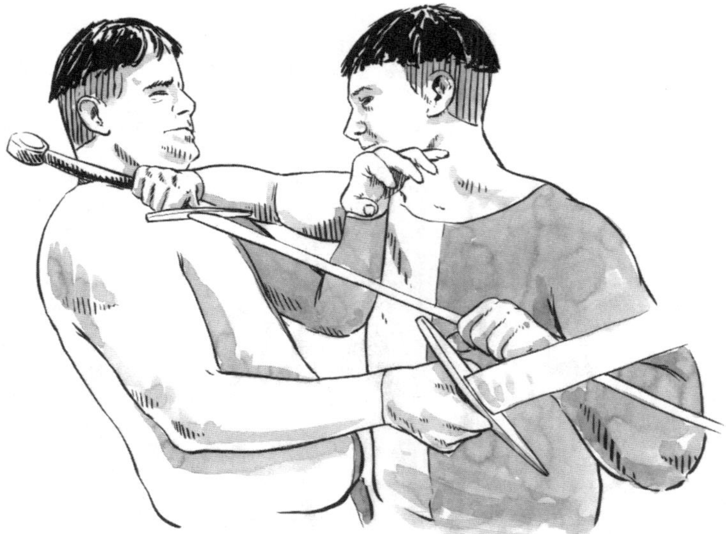

Nachrayßen

Nachrayßen means "traveling after." By moving after your opponent's action you exploit his opening and counter-time him, giving him less space in which to defend and attack. By attacking him in the initial phase of his own attack you catch him at the very worst moment (from his point of view). He is committed and will have great difficulty altering his movement while in the middle of it. The *nachrayßen* can be applied to all situations in fencing and as such it is a principle and not one or two techniques to be executed in special situations. *Nachrayßen* works best when you use the blade; it is possible to step in to wrestle, but Ringeck advises us to travel after and strike if our opponent strikes and thrust if he thrusts. The attack must then be done in such a way that you are certain to gain the advantage. Again we see the concepts of *vor* and *nach*. The attack should be placed in the same opening just as the opponent's weapon moves foward. (If, for example, he is pulling the sword back to the high left to strike from above, you strike at that window.) But you do not necessarily aim at his weapon; what you want to do is catch him in a bad position. It is important to lead with your weapon and follow with feet and body (as in all movement) since you want to thrust your sword at your opponent as fast as possible in order to regain the initiative. Often you will not score a finishing hit with the *nachrayßen*, but you will create the opening for a follow-up. And in many situations it is good to alternate strike and thrust: that is, you travel after with a strike and finish with a thrust or cut.

"Learn two nachrayßen *or cut in the weapon."*

There are two kinds of *nachrayßen* and you will learn them both. Use the first against an *oberhaw*: when he lifts his sword to strike travel after him with a strike or a thrust and hit him in the upper opening before he can make his strike. Or fall with your long edge from above on his arm and thereby push him away from you.

"Nach raysse lere zwyfach oder schnyd in die were"

Das ist das dü die nach raysen wol lernen solt / wan sy sind zwayerlay die ersten tryb / wen er dir oben in will hawen so merck die will er das schwert uff zücht zu dem schlag so rayse im nach mitt dem hawe oder mitt ainem stich und triff in zu den obern blöß ee er mitt dem haw wider kump oder fall im mitt der langen schniden oben in sin arm und truck in domitt von dir

The idea is to strike at your opponent's upper opening where his sword is, since he will find it very hard to defend himself against it. You might also consider aiming at his hands because they are very easy targets compared with his head.

It is important that you try to strike him as he moves into a new position. Then it will become very difficult for him to parry your blows.

You can also place your sword on top of his arms if he is moving his sword up from a lower position.

When you have struck him, place your sword over his arms, You can then either take another step with your lead foot or make a pass with your rear foot as you push him back.

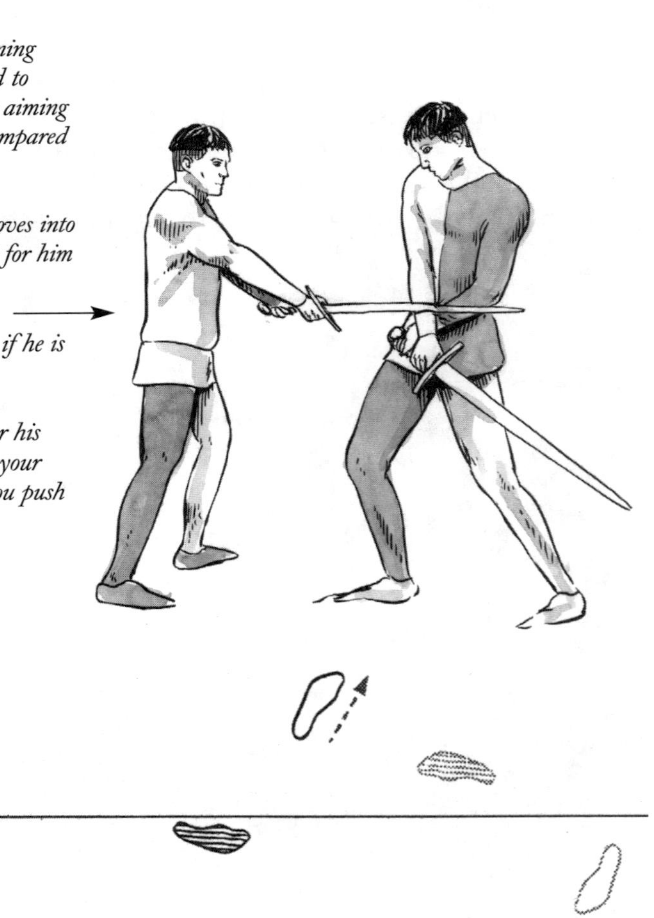

"Another nachrayßen.*"*

When he strikes with an *oberhaw* and in the strike lowers his sword toward the ground, then travel after him with a strike to the head before he has time to lift his sword again. Or if he wants to make a thrust at you and pulls his sword back, then travel after him and thrust at him before he has time to do it himself.

"Aber ain nachraisen"

Item wann er dir von oben zu hawet laß er dann sein schwert mitt dem haw nider gen zu der erden so rayß im nach mitt ainem haw oben ein zu dem kopfe ee er mitt dem schwert uff kumpt Oder will er dich stechen so mörck die wil er das schwert zu im zücht zu dem stich so rayß im nach und stich in ee wan er sinen stich vol bringt

Strike the schaytelhaw *as before with your arms high. Beware that he can make a low thrust at you.*

As he pulls back, step forth and thrust. Make the step with your lead leg if possible–it is faster.

"Two outer takings, your work begins there. Test the bind, if it is weak or hard."

"Zwai eußren nym dein arbait dar nach beginn und brieffe die gefert / ob sy sind waych oder hort"

Note: there are two outer takings (*eußern nym*), that is, two *nachrayßen* at the sword. Do them like this: when he fails in front of you, then *nachrayße* after him. If he deflects this, remain with your sword on his and discern if he is weak or strong in the bind. If he then lifts his sword with strength, then guide your blade over and to the outside of his sword and thrust at his lower openings.

Merck die zwi eussere nym das sind zwai nachraisen am schwert das trib also wan er sich vor dir verhawet so rayß im nach versetzt er dir den das so plyb mit dem schwert an dem sinen und brieff / ob er mit denn gefert waich oder hert ist hept dan er mit dem schwert das din mit sterk ubersich so lang din schwert ussen yber das sin und stich in zu den underen bloß

Your opponent can fail to hit you for many reasons: on purpose, because you step back, and so forth. Strike at once at his upper opening.

If he lifts his sword to deflect your blade from below (as seen at right) or with the point up, feel his intent in the bind. Note that the opponent can strike with either long or short edge (it does not matter), but try to avoid an edge-on-edge collision. (In illustrations, the correct angle of the sword can't always be shown since you would often see only a narrow line).

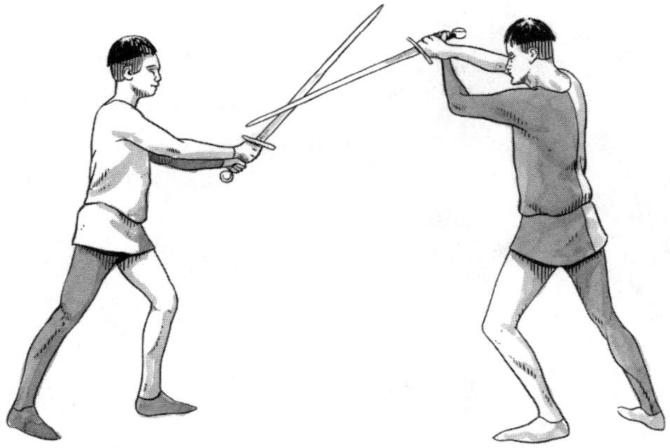

Wind your blade over his to the outside (this is a mutieren*) and make a thrust at his lower opening. In the winding you will push his sword to the side or simply thrust at him over it. Then his sword will be close to your body, so you must act fast and move past it immediately after the thrust. He cannot press against you because you have the leverage.*

Step forward with the lead foot as you thrust. Use your body in the thrust. The thrust can be done from above or once you clear his sword horizontally.

This also works if he lifts with the short edge up and binds with strength.

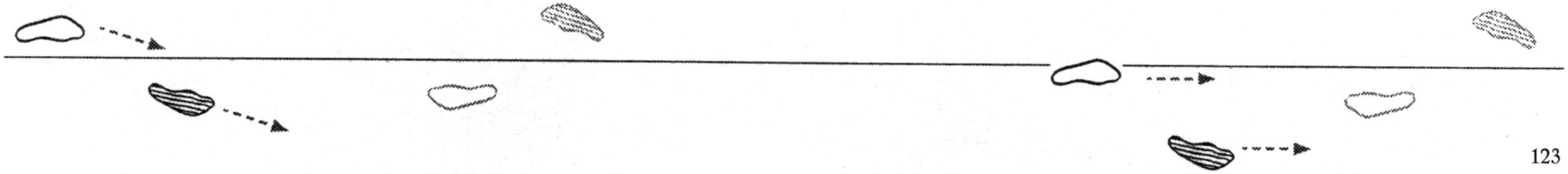

"The other outside taking."

Also, when you fence with him using *unterhaw* or other techniques, or when he forestalls you and winds his sword from above against yours so that you cannot raise your sword, then hold strongly with your sword under his. When he then works toward your upper openings, follow with your sword and capture the weakness of his sword with your long edge, then push him down and stab him in the face.

"Die ander eußer nym"

Item wen du geim fichtest uß under hawen oder sunst uß den underen geferten uberylet er dich den und windt dir uff da schwert / ee / wan dü mit uff komest so blieb mit dem schwert starck unden an dem sinen wend er den und arbait dir zu dem obern bloß so volg mit dem schwert nach und mit der langen schnyden nym im die schwech sines schwerts und truck nyder und stich im zu dem gesicht

Your unterhaw *is wound against and pushed down; bind with strength and wait for the moment when your opponent begins to work at you.*

Step forward with your rear foot and turn your sword over from your right to left of your middle line. Capture his weak with the strong of your sword in this movement. The point will then be aimed at his face.

Take a step with the lead foot and thrust at his face. This works on both sides and is very useful.

"Learn the fühlen; indes *is a word that cuts sharply."*

You must learn and understand both the concept of feeling (*fulen*) and the concept of simultaneity (*indes*), since they are connected and form the highest level in the art of fencing. Do it like this: when you bind the other by the sword, then you must in the very moment that the blades touch feel (*fulen*) if he has bound strongly or weakly. As soon as you have discerned this think of the word simultaneously (*indes*). That is, you shall at once and skillfully work with your sword, weak or strong, against the next opening as you see fit. Thus, he will be defeated before he knows it.

You shall also remember the word simultaneously (*indes*) at all the binds at the sword, for simultaneously (*indes*) doubles (*dupliert*) and *indes* changes (*mutiert*), *indes* leaps through (*durchlauffen*) and *indes* takes the cut and *indes* wrestles (*ringt*), and *indes* takes the sword (*schwertnehmen*) to disarm him. *Indes* does in fencing whatever your heart desires. *Indes* is a sharp word that cuts all fencers that do not know it. And *indes* is the key with which the art of fencing is unlocked.

"Das fulen lere Indes das wort schnydet sere"

Das ist das dü das fulen / und das wort indes wol lernen / und verstan solt wan die zwy ding gehören zu samen und sind die größte kunste zu dem fechten und das vernym also wen ainer dem anderen an das schwert bindt / so soltü in dem alß de schwert zu samen glutß schon zu hand fullen / ober waych oder hort angebunden hab / und alß bald dü das empfundest haust so gedenck an das wort / indes / das ist das dü in den selben enpfinden behendtlichen nach der waich und nach der hort mit dem schwert solt arbaiten zu der nechsten bloß so wirt er geschlagen ee wann er sin gewar wirt

Item du solt an das wort / indes / in allen an binden des schwertz gedencken wann / indes / dupliert und / indes / muttiert / indes / laufftdurch und / indes / nimpt den schnitt / indes / ringt mitt / indes / nimpt das schwert / indes / tut in der kunst waß dein hertz begert / indes / ist ain scharpffes wort damit alle fechter verschnitten werden by des worts nichten wissen Und das wort / indes / ist och der schlissele damit alle kunst des fechtens uffgeschlossen wirt

The two concepts of indes *and* fulen *are as shown in the text simple enough to understand but very difficult to master. There is a tendency with many fencers to stop once they have made contact with the opponent's blade or body. Compare this with modern sport fencing and you see that they rely on touch, feeling through the blade what the opponent intends to do. And that is the idea here, to use the sword as an antenna so that you grasp the opponent's intentions in an instant. It is difficult to practice this with sword in hand, so start out by using only your hands and bodies. The hardest part is that if you use a cognitive thought process to divine your opponent's intentions, you lose. You must know without thinking what he intends to do, and before he knows it himself.* Fulen *is the intuitive knowledge and* indes *is the moment without rhythm, when you are free to act. One tells you when to do it, the other what to do.*

"Traveling after twice, hit well with the old cut."

If he fails with a strike in front of you, then travel after with a strike to the upper opening. If he then moves up and winds against your sword from below, then note: as soon as the swords clash, hit on his arm with the long edge and push him away from you. Or cut him across the face. Practice this on both sides.

"Nachreisen zwifach / trifft man den alten schnitt mitt macht"

Das ist wann er sich vor dir verhawet so rayse im nach mitt ainem hawe zu der obern blöß fört er dann uff und windet dir unden an das schwert so merck aber so bald ain schwert uff das ander blitzscht so val im vom schwert mitt der langen schnyden über sein arm und truck in also von dir Oder schnyd in vom schwert durchs maul das tryb zu beden sytten

If you catch his sword lower down, you may instead cut him on the upper side of his forearms or across his biceps with a forward cut. But, often the impact of his strike from below will lift your arms too high for that and you must cut from below instead.

This will protect you from his blade because he cannot come around without being cut.

Or, from the bind, in time with a forward step cut at his face or arm.

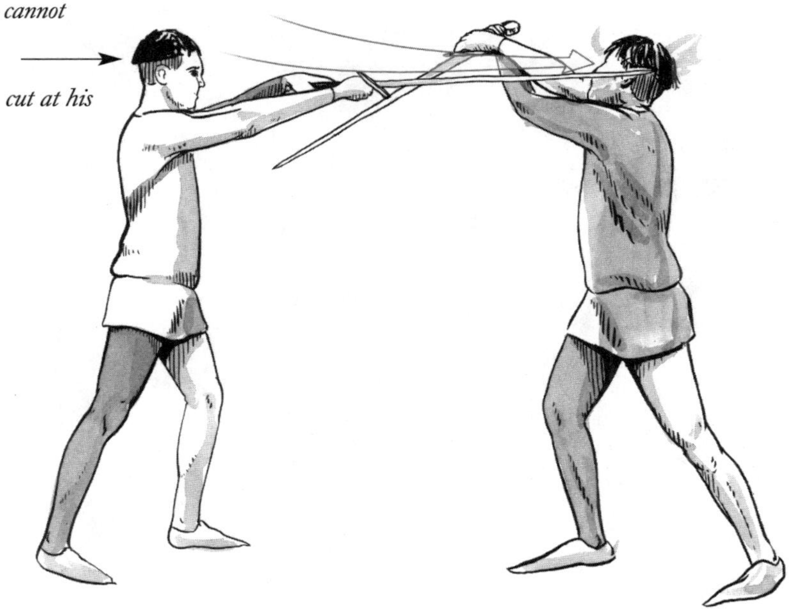

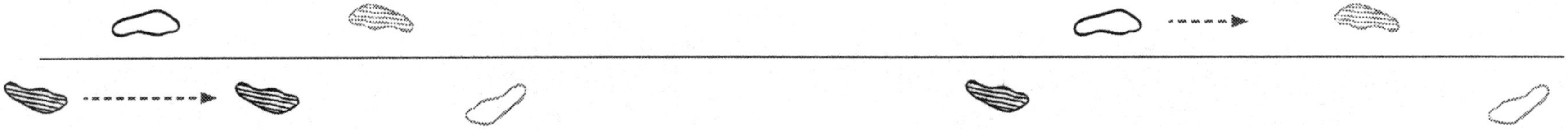

Absetzen and Durchwechseln

Durchwechseln is the term for changing the angle of the straight line that connects you, from the center of your body, to your opponent's body. Sidestepping is the less glamorous word for it. By "changing through" you can find new openings and lines along which you may attack. This is usually done with a *winden,* for example.

By winding the sword you change the angle from the tip of your sword to its target, and the following thrust comes as a natural extension of the movement. It is not simply dropping your sword under his, or disengaging from a bind. That is merely the application of the principle. In the *durchwechseln,* there is movement of the feet, by necessity, but the weapon must still lead the movement. You initiate with the sword and the body follows. In practical fencing with longswords, you cannot change the angle between the two of you more than 45 degrees or so. It is not effective to try to increase that angle since your opponent has to make a smaller movement to counter yours. The principal *durchwechseln* consists of altering the angle of the sword, and then shifting the body slightly to that side as you press your attack from the new angle. Most effective is a "changing through" followed by a thrust.

Absetzen means to set aside. You can set aside a thrust or a strike, both more or less with the same movement, starting from the *pflug* or the *ochs.* It is a deflecting motion against a thrust where you use the flat of the blade to steer the opponent's point aside and hit with your own. You catch strikes on your blade, using the flat; his blade then slides down to the crossguard and you can counter with a thrust. However, it is very difficult to use the *absetzen* against a physically strong fencer: if he is positioned well, he will knock your intercepting sword back at you. All in all, there are four basic *absetzen,* two upper and two lower, that correspond to the positions of the *ochs* and the *pflug. Absetzen* as such can be said to be moving between *ochs* and *pflug* and deflecting an incoming attack on the way.

"He who wishes to strike below, overrun, and he will be shamed. When he strikes from above, then strengthen what I laud and do your work, or press hard twofold."

If he aims a strike or a thrust at your lower opening while closing in, do not deflect this. Instead wait until you can run him over (*überlauffen*) with a strike to the head or with a high thrust. Thus you will defeat him, since the upper strikes and thrusts all have a longer reach than the lower ones.

"Wer unden remet uberlauffe den / der wirt beschemet wenn klutzscht oben / so störck das will ich loben dain arbait mach / oder hert druck zwifach"

Das ist wann er dir in dem zufechten mitt ainem haw oder mitt ain stich remet der undern bloß das soltu im nicht versetzen sonder wart das du in überlauffest mitt ainem haw obe ein zu dem kopffe oder im den ort oben ansetzest so wirt er von dir beschemet wann alle ober haw und alle ober ansetzen überlangen die undern

As he begins to strike low, strike high. You may even lift your grip so that the sword is almost horizontal to strike faster. If you make a high thrust, remember that even though it is faster than a low one, a thrust is sometimes slower than a strike.

You can also simply aim at his hands. High does not by necessity mean "at the head." Consider anything above his blade as "high." Aiming for his hands is a very effective counter, hard to avoid.
It is possible to hit both his hands and head at the same time. It all depends on distance and angle.

"Learn absetzen. *Strike and thrust with skill when he thrusts so that your point hits and his misses, on both sides you will strike all with a step."*

You will learn to deflect strikes and thrusts with skill so that his point is broken and yours hits. Do it like this: if he stands before you as if he is about to thrust at your lower opening, assume the guard of the *pflug* on your right side and expose your left. If he then thrusts at your (left) opening, wind your sword against his toward your left side and take a step toward him with your right foot. Your point will hit and his will miss.

"Lern absetzen / haw stich kunstlich letzen wer uff dich sticht / das din ort trifft und sinen prücht von baiden sytten / triff alle mal wiltu schrytten"

Das ist das du lernen solt hew und stich also mitt kunst *absetzen* das dein ort in treffe und im der sin gebrochen werde. Und das vernym also wenn ainer gegen dir steet und helt sein schwert alßer dir unden zu stechen wölle so stee wider gegen im in der hut des pflugs von diner rechte sytten und gib dich blöß mitt der lincken stich er denn dir unden zu der selbigen blöß so wind mitt dem schwert gegen synen stich uff din lincke sytten und schryt zu im mitt dem rechten fuß so trifft din ort und der sin fält

As you begin to shift your feet, you move your sword toward your left. This will deflect his incoming blade. As you parry, do not let your sword point drift outward: keep it pointed at your opponent and remember to keep your elbows close to your body.

As you complete the movement, extend your arms slightly as you complete the step, thus putting the weight of your body behind the thrust. Try to turn your sword as you make the deflection so that your wrists turn slightly upward as you meet his blade (the long edge will be on top). This gives you stability, and you can keep the point aimed at your opponent.

"Another technique from the absetzen."

If you face him in the guard of the *pflug* on your left side and he strikes at your upper left opening, then move up with the sword to your left side with the crossguard in front of your head. Take a step toward him with your right foot and stab him in the face.

"Aber ain stuck von absetzen"

Item wann du gen im stast in der hut de pflugs von diner lincken sytten hawt er dir dann zu der obern blöß diner lincken sytten so far uff mitt dem schwert unnd uff die lincken sytten gegen sinen haw / das gehültz für din haupt und schryt zu im mitt dem rechten fuß und stich im in das gesicht

As your opponent strikes, simply lift the crossguard of the sword and catch his strike on your blade at the strong part. Use the flat or angle the sword up and get the same contact angle as with a zornhaw *parry, which is stronger but does not offer a good opening to strike at once. His blade will then slide down to your crossguard or bounce off, depending on the angle of his sword and the force used. You do not catch his strike with the crossguard. Note that if you don't lift your sword so that your point is up, there is a risk that he will strike your front hand over your crossguard. It is possible to lower the point somewhat compared to the illustration.*

You can do this technique in several ways, leading with either foot. As you thrust you can make a pass or a step as the situation demands. The principle is to deflect the strike, lower the point, and thrust straight forward at his face.

Lower your point by pushing the pommel slightly down and keeping the cross-guard steady. Step forward with the lead foot and extend your arms to stab him in the face. As you catch the strike, raise yourself a bit from the low pflug *position–this gives you strength.*

It is also possible to lift to a left ochs *in a circular movement that will deflect an incoming strike and leave you in a good position for a thrust. Explore all options.*

"Learn durchwechseln *on both sides and thrust with intent. If he binds with you, change through and find him open."*

You will also learn *durchwechseln.* If you strike or thrust at him and he strikes against you or parries and binds your sword, then let the point of your sword glide under his to the other side and thrust at him on the other side–there you will find an opening.

"Durchwechseln lere von bayden sytten stich mitt sere wer uff dich bindet durchwechse in schier findet"

Daß ist daß du das durchwechseln wol lernen solt und das tryb also wenn du im in dem zufechten ein hawest oder zu stichst will er dir denn mitt ainem hawe oder mitt versetzen an daß schwert binden so lauß den ort under sinem schwert durch wischen und stich im do mitt sere eyn zu der anderen sytten so fündest du an im schier de blöß

*After the bind, as you feel (*fühlen*) that there is a moment's opening, drop the point of your sword under his as you step and wind to the other side. It is possible to wind both high to the* ochs *and low to the* pflug*. Thrust at his face at once.*

You can also change through at other times by finding the opening where you can step and wind the sword into a thrust, or at times into a strike as the situation demands.

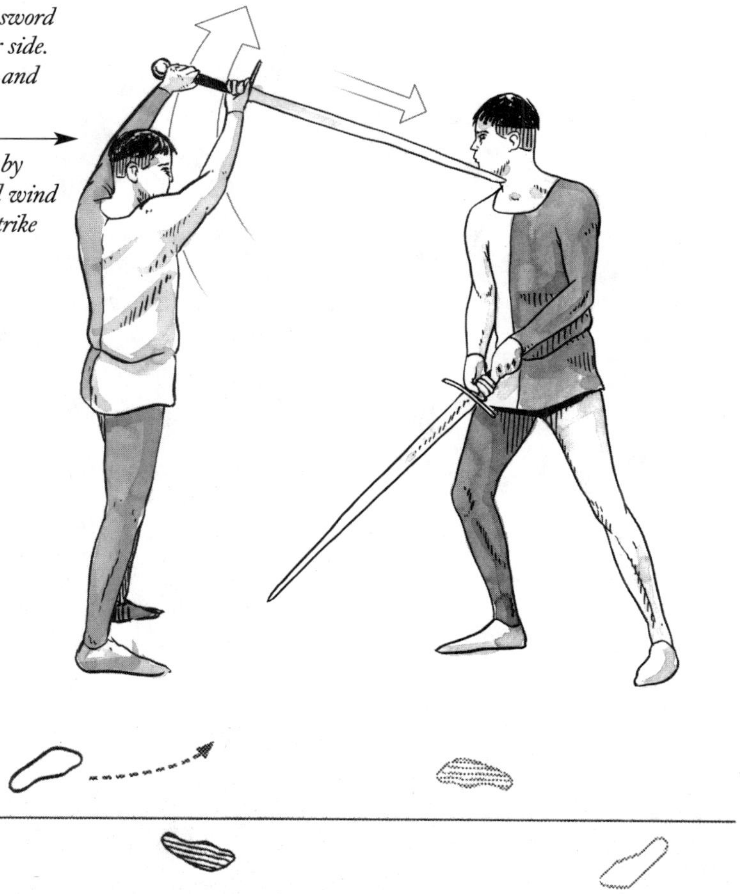

"Step close in the bind, find where it is good to jerk. Jerk; if you hit, then jerk again. Find the action and jerk in all contacts; then you will emulate the master."

When you close with him, strike from the right shoulder at his head. If he then binds your sword with a deflection or other technique, then step close while in the bind and lift your sword swiftly up and away from his and strike him at the other side of his head. If he deflects this again, strike once more at the other side and work skillfully at the upper openings using *duplieren* and other techniques.

"Tritt nahend in binden daß zucken git gut fünden zuck trifft er zuck mer Arbeit / erfinde das tut we Zuck in allen treffen / den maistern wiltu sy effen"

Das ist wenn du mitt dem zufechten zu im kumst so haw von der rechten achseln starck oben ein zu dem kopff bindet er dir dann mitt versetzen oder sunst an das schwert so tritt im bande nahent zu im und zuck dein schwert oben abe von dem sinen und haw im zu der andern sytten wider oben ein zu dem kopff versetzt er dir das zu dem andern male so schlach wider zu der andern sytten oben ein und arbait behentlich nach der obern blöß die dir werden mag mitt dem duplieren und mitt andern stucken

Zucken *is basically continual striking on alternate sides. What is important to remember is that you must use more than strikes (that is,* duplieren*), cuts, and even thrusts. Especially in the last attack, the more you surprise him the likelier you are to hit.*

Strike at his head, alternating sides. After the third strike you can usually hit him with a thrust at the side other than the last tried. Or you can feint and, on the third or fourth strike, aim at his knee. Stepping with the strikes adds power, but makes you slower in the sense that your strike must land when you are balanced, so you will never be able to strike faster than you can move. And you can move your arms faster than your body. That said, it is important to not stay still; you should move to generate acceleration in the attack or to avoid your opponent. 1he easiest way to generate force is to use the waist and hips.

It does not matter what kind of strike you make from the right side, but remember that some are easier than others to change from. Instead of a zwerch, *you could strike* zorn-haw *or alternate between them. Also remember that you can change between high and low strikes.*

Ringen am Schwert and Durchlauffen

Close-in fighting is important in Ringeck's and Lichtenauer's system and is well represented by the wrestling and *durchwechseln* techniques. In essence, these consist of the same elements, namely evading the force of your opponent by avoiding him and using leverage to push or throw him. The instant of unbalancing your opponent is the most crucial in these techniques, as in all that rely on closing the distance and grappling or wrestling with your opponent. If he retains his balance, you will be at a distinct disadvantage and in danger of being thrown or disarmed yourself. It is very easy to see that the close techniques delivered at the closest of the three distances are directly derived from true wrestling, and are only adapted for use with a longsword in your hands. In all wrestling there are a few key points to remember. You must move in close and disrupt your opponent's balance; it is impossible to do any wrestling with success without disrupting his balance. Second, you must act with determination because moving in to wrestle is dangerous. Also remember misdirection–imagining that you can get a hold on a person "just like that" is fantasy. You must distract him on the way in, punch him in the face, kick him in the groin, elbow, or headbutt him in the face, and so forth. This will create the space needed to apply the technique. When you push, throw, or place a lock on your opponent, do so with force by using the body and not just the arms. The crucial moment is when moving in; that is the really dangerous ground and must be covered with speed and intent.

Ringen am schwert *means wrestling with the sword, and* ringen *is often used alone to indicate the use of wrestling moves.* Durchlauffen *means to run through (in this case an opening in your opponent's defenses).*

"Leaping through, let (the sword) hang by the pommel, wrestle if you will. Whoever comes at you with strength, remember to leap through."

Note: If someone rushes in with his arms over his head and tries to overwhelm you from above with strength, then lift your arms high also and hold your sword with the left hand on the pommel above your head and let the blade hang down over your back. Duck under his right arm and leap with your right foot behind his right. As you leap, grab him around the body using your right arm, lift him on your right hip, and throw him (down) in front of you.

"Durchlauff lauß hangen mitt dem knopffe gryff wiltu rangen wer gegen dir störcke durchlauff damitt mörcke"

Mörck das ist wann ainer dem andern einlauffe fört er dann uff mitt den armen und will dich oben mitt störck überdringen so far och uff mitt den armen und halt din schwert mitt der lincken hand by dem knopff ober dinem haupt und lauß die klingen hinden uber deinen ruck hangen und lauff mitt dinem haupt durch under sinem rechten arm und spring mitt dem rechten fuß hinder sinen rechten / und mitt dem sprünge far im mitt dem rechten arm vornen wol um den lyb und fasse in also uff dem rechten hüffe und wirff in für dich

If you begin the technique from the bind, lift his blade and push it to the side as you step in. Lift it over your head, then let go with your right hand as soon as the crossguard is in front of your head, and continue the movement using the grip with your left hand on the pommel. The stepping is done at about 45 degrees forward, but don't step too far out or you will not be able to grab him properly.

As you get close to him, let the sword hang down diagonally over your back. Hold it by the pommel with your left hand. Do not let go of the sword with your right hand and expect to be able to hold against his strike. The idea is to step so close that his sword will, at worst, slide off your blade without much force.

Grab him with force around the chest, not the waist; you can push your arm up so that you press against his armpit. There must be no space between you: you must be close to get a good grip. At right is the position after stepping in and grabbing him.

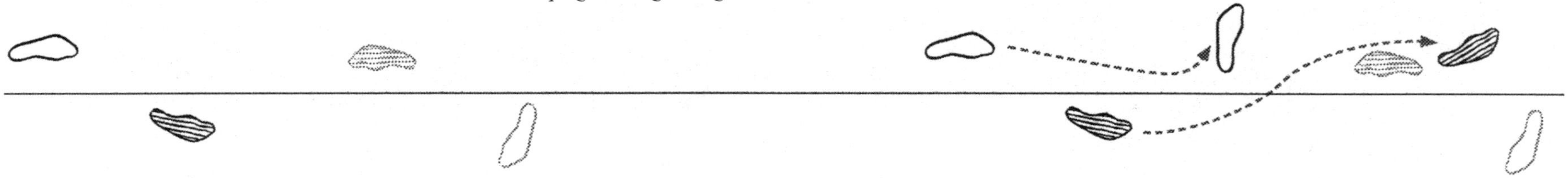

As you throw him, turn to your left and straighten your right leg as you heave him over. If you don't, the throw is less effective. Remember that you must use your body to make an effective throw. It is the twist of the waist and the pull that takes him over.

Try to throw him away from you and down; then at once grab your sword with both hands and strike him in the head or upper body. The technique must be done with both speed and force to be both understood and correctly executed.

"Another durchlauffen."

When he tries to rush in to overwhelm you with the sword high and using strength, hold your sword with your left hand on the pommel and let the blade hang down over your back. Duck under his right arm but let your right foot remain in front of his right. Grab him then with the right arm behind his back and throw him behind you over your right hip.

"Aber ain durchlauffen"

Item wann er dich in dem einlauffen mitt dem schwert oben mitt störck über dringen will so halt dein schwert mitt der lincken hand by dem knopff und lauß die clingen über deinen ruck hangen und lauff mitt dem häupt durch under sinem rechten arm und blib mitt dem rechten fuß vor sinem rechten / und far mitt dem rechten arme hinden wol im umb den lyb und fasß in uff din rechte hüff und wirff in hinder dich

The principle is the same as in the last technique: sidestep him as he moves in, protecting your back with your blade held in your left hand. You can also do all these techniques if he has bound your sword with a high kron.

It is important that you step in close, otherwise you will not be able to reach him. There should be no space between the two of you, otherwise he can easily slip out of the technique by stepping around you. Lean forward in the throw and retain good balance, otherwise he can throw you.

As you pull him down with your arm, lean forward with your body and lift/kick your right leg forward and up, attacking his primary point of balance here, his front leg.

"A wrestling with sword."

If he rushes in, let go of your sword with your left hand and hold it in your right. Push his sword away from you to your right side using your crossguard. Leap with your left foot, putting it in front of his right. Grab him around the body with your left arm, lift him on your left hip, and throw him down on the ground in front of you. But be careful that you do not fail.

"Ain ringen am schwert"

Item wann ainer dem andern einlaufft so lauß dein schwert uß der lincken hand und halt es in der rechten und stoß sein schwert mitt dinem gehültze uff din rechte sytten von dir und spring mitt dem lincken fuß für sinen rechten und far im mitt dem lincken arm hinden wol um den lyb und fasse in uff din lincke hüffe und wirff in für dich Doch lug das es dir nit fel

This technique can begin like the last one, from a bind, or if he tries to strike you from above. Step in close as you push his sword away. There should be no space between you; this may be hard to do, but step in aggressively.

Be sure that the point of your sword is aimed upward. If not, your opponent's sword will escape its capture at your crossguard. Pull him over your leg and down. Use your leg to trip him by pushing it back at the same time as you pull him over it.

Turn your body in the throw so that you remain with your centerline aimed at your opponent, ready to strike or thrust.

"Another wrestling with sword."

If he rushes in, let go of your sword with your left hand and hold it in your right. Push his sword away from you to your right side using your crossguard. Leap with the left foot behind his right and grab him with your left arm around the breast and throw him back over your leg.

"Aber ain ringen am schwert"

Item wann ainer dem andern einlaufft so lauß din schwert uß der lincken hand und halt es in der rechten und stoß mitt dem gehültz sein schwert uff din rechte sytten von dir und spring mitt dem lincken fuß hinder seinen rechten und far im mitt dem lincken arme vornen under sin brüste wol um den lyb und wirff in uber dein fuß hindersich

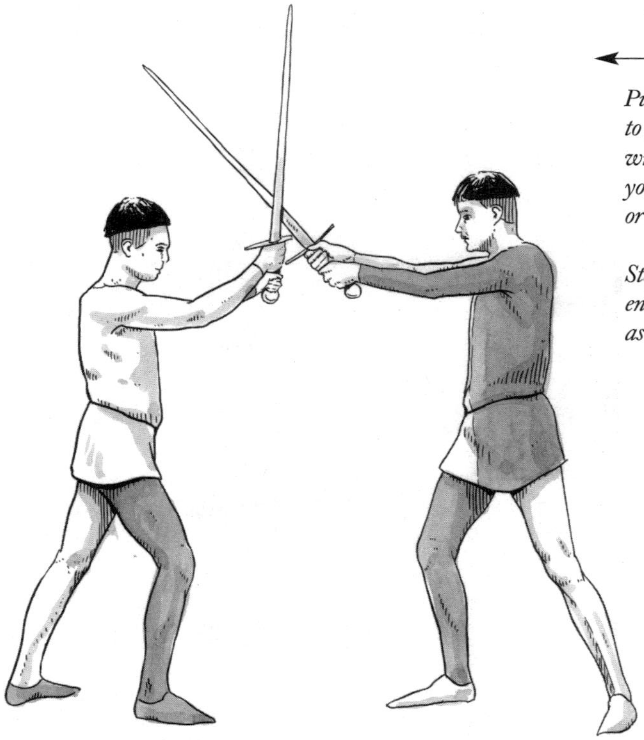

Push his sword to the side, using your waist as you turn to generate force. (Do not try to move his blade aside with your arms only.) As you turn to your right, shove your left arm in as high as possible over his chest, above or below his arms.

Step in with the left foot. Make sure that you step far enough so that he cannot simply step back over your foot as you push him over your leg.

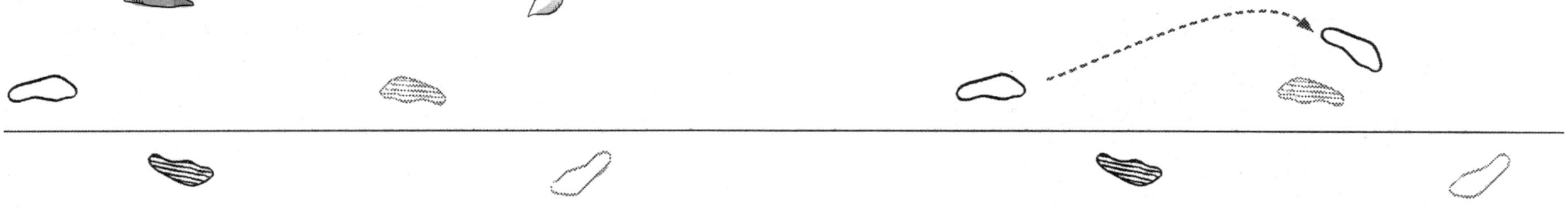

Push him back over your leg using your waist and body, not just your arm. You can kick his knee aside as well, or even break it at the joint as you throw him.

It is more efficient to grab his head and pull it back than it is to push his upper body. You can also strike him in the throat as you position your arm, or use your elbow. If he is wearing a helmet, you can push or pull it by grabbing its rim. If he has long hair, grab it. And remember to use force; without it no wrestling will work.

"Another wrestling with sword."

If he rushes in, let go of your sword with your left hand and hold it in the right. Lift the pommel up and over his right arm and then pull down. Grab his right elbow with your left hand and leap with your left foot in front of his right and push him over your leg to your right side.

"Aber ain ringen im schwert"

Item wann du mitt ainem eynlauffest so lauß dein schwert us der lincken hannd und halt es in der rechten und far im mitt den knopff ussen über seinen rechten arm unnd züch domitt undersich und mitt der lincken hannd begryff sein rechte elnbogen und spring mitt dem lincken fuß für seinen rechten und ruck in also über den fuß uff din rechte sytten

This is one of many ways to engage in this technique. To do this well, the breaking must be done with a fast and determined motion.
Lift the pommel up and slightly to the right, giving you space to move in close.

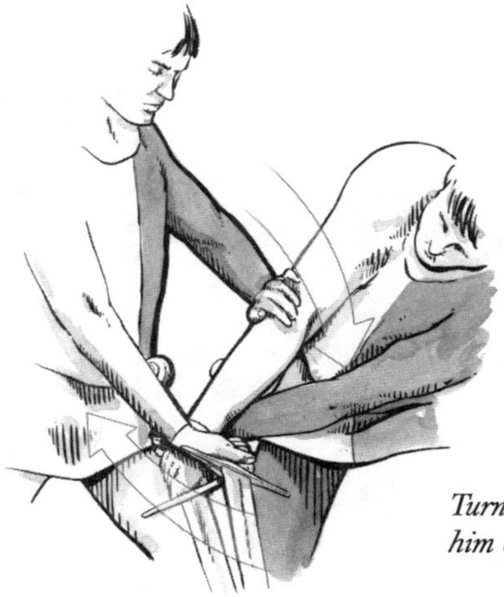

Turn at the waist as you push him over to generate force.

Step in with your left foot in front of his and simultaneously slam the pommel down between his hands. Grab his elbow. Then pull your sword toward you and at the same time push his elbow in the opposite direction. Make sure that you have locked the joint.

You can take a small step to the side and rear with your hindmost foot to get it out of the way as he falls.

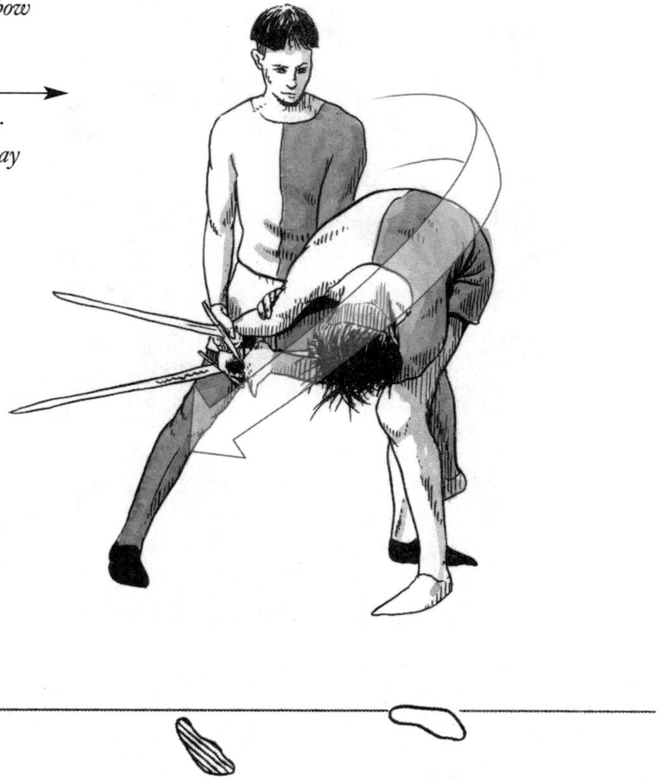

"Another wrestling with sword."

If someone rushes in, then move your left hand over his right arm and grab hold of your own right arm. Push his right arm, using your own, over your left and leap with your right foot behind his right, then turn to your left side. Thus you will throw him over the right hip.

"Aber ain ringen im schwert"

Item wenn ainer dem andern eynlaufft so far mitt lincker verkörter hand über sinen rechten arm und begryff da mitt dinen rechten arm und druck mitt dine rechten arm sinen rechten arm über dinen lincken und spring mitt dinem rechten fuß hinder sinen rechten und wende dich von im uff dine lincke sytten so wirffestu in über din rechten hüffe

This technique is somewhat complicated but quite fun to perform.

From the bind move in and, as you step, push his sword to your right with the crossguard, making sure that you keep your point up; do not let it drop. Move your left hand over his right arm, palm down, up between his hands and move it to your right arm.

Grab your right arm with your left hand. The swords will be placed on your opponent's neck or upper arm.

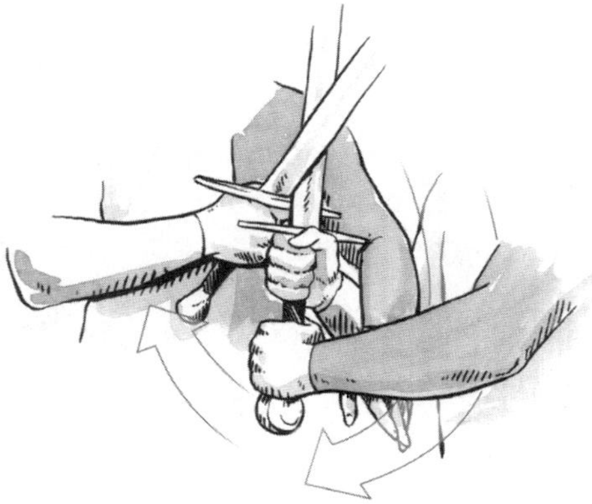

Grip with force and hold him tightly. As you move your hand over his arm, press down on his right elbow to weaken him. Use the weight of your body by sinking slightly at the knees.

Move your left hand up and place it on top of your right forearm. You will now be able to press his right hand between the grip and your left forearm. He will not be able to move.

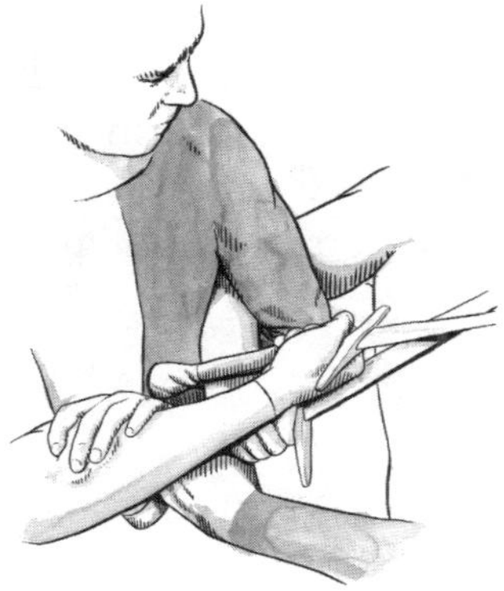

Make sure that your left foot is not too close to his right or you will not be able to step in with your right foot.

Step with your right foot behind his right, place it firmly, and turn to your left at your waist at the same time. Do this with force.
Practice this technique with care because one can take a very hard fall.

"Another wrestling with sword."

If someone rushes in at you, then turn your left hand around and move it over his right arm. Grab the hilt of his sword between his hands and pull it to your left side. Thus you will take his sword away and it will not go well for him.

"Ain ander ringen im schwert"

Item wann ainer dem andern eynlaufft so verkör dein lincke hand unnd far domitt uber sinen rechten arm und begriffe domitt syn schwert by der handhäbe zwischen sinen baiden henden und rucke uff dein lincke sytten so nimest du im das schwert

Step in close and move your left arm over his right. Push down on his elbow with your arm to weaken his grip and break his balance. Take hold of the hilt of his sword, preferably between his hands.

It works just as well from the other side, and even with the other foot leading.

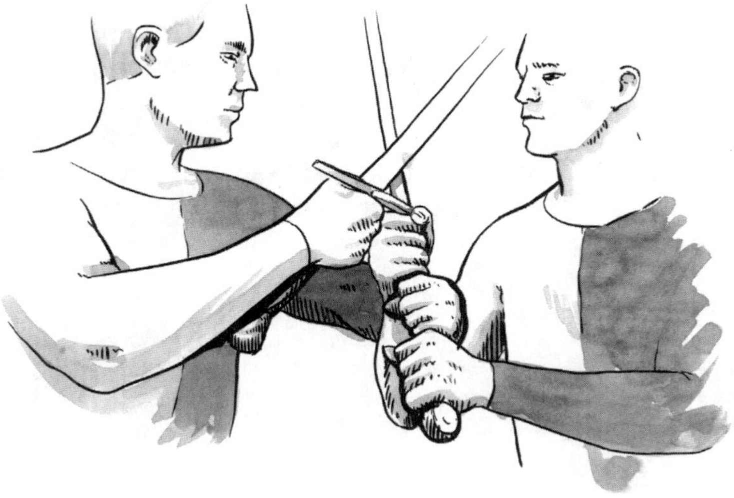

It is important that you press down on his elbow with your arm, and that you keep your sword as a guard between the two of you. If not, you will get cut in the face.

Step back with your left leg and turn at the waist, retaining a strong grip on his sword. You will twist his hands and break his balance. This must be done with speed and determination. Here you can also kick to his groin or knee if he does not let go.

Pull his sword up and toward you in time with stepping back with the left foot.

"A disarming of the sword."

If he binds your sword by deflections or other techniques, use your left inverted hand to take hold of both swords at the middle of the blades. Holding them together tightly, move your right hand to your left side with the pommel down, move below and through both his hands, then jerk up to your right side. You will take both swords this way.

"Ain schwert nemen"

Item wenn er mitt versetzen oder sunst an din schwert bindet so begryff mitt lincker verkörter hand bayde schwert mitten in den clingen und halt sy veste zu samen / und far mitt der rechten hand gegen diner lincken sytten mitt dem knopff unden durch im über sein bayde hende und ruck domitt übersich uff din rechte sytten so belyben dir bayde schwert

From the bind, step forward quickly to grab the blades, or step forward as you tear the swords to your left side, using the momentum of your body to add force.

Push through under his hands, stepping in with your left foot as you do so. Continue pushing up and over his hands. This will twist his hands and force him to let go. As you begin to move back to your right side, step back to add momentum. After stepping back, use this momentum to pull your arms to your right side.

Your body will take his sword, not your arms. They are too weak, so use them only to grip his sword as your body takes it away.

Das Schnitt

The cuts, or *schnitt*, can be done in several ways, both as a drawing cut and as a pushing cut. The drawing cut is excellent for continuing the movement back and around to land a strike; the pushing cut places you at a distance where wrestling techniques may be employed and you have taken the initiative from your opponent by placing cutting. The cut is perhaps the best way of using the *nachrayßen* if your opponent lifts high to strike or tries to run in. By stepping in and closing the distance and simultaneously positioning the sword to cut, you both close his line of attack effectively and counter it. It is always possible from the bind to make one cut or another; this often implies the use of *winden* to succeed. It is essential to fully understand that cutting is close-in work–you make the best cuts when you are close. It is also important to understand that the *schnitt* and *durchwechseln* belong together, by "stepping through" and cutting you can turn a losing position into a winning one. Pressing and pushing the opponent in any way is also using the *schnitt.*

Schnitt *means making cuts in any manner rather than strikes or thrusts.*

"Cut off the hard from below and in both bind; there are four cuts, two below and two above."

Note: There are four cuts. Do the first one like this: If someone rushes at you with arms up and tries to push through forcefully at your upper left, wind your sword and fall on his arm below the crossguard with your own arms crossed and using the long edge, and push him upward. If he leaps at you on your right side, strike with the short edge on his arm and push him upward as before.

"Schnyde ab die hörten / von unden in bayden geferten vier sind der schnitt / zwen unden zwen oben mit"

Mörck es sind vier schnitt den ersten tryb also wann er dir einlaufft und hauch uff fört mitt den armen und dich oben gegen diner lincken syten überdringen will mitt störck so verwend din schwert und fall im mitt der langen schniden mitt gecreutzten henden unnder sinem gehültze in sin arm und truck mitt dem schnitt übersich Oder laufft er dir ein gegen diner rechten sytten so fall im mitt der kurtzen schniden in sin arm und truck übersich alß vor

This can be done in several ways, using either edge. The idea is to meet him instead of waiting, close the distance, and place your sword in position for an upward directed cut. If you remain still you will not reach him and he will strike you instead.

Step forward and to the side as you wind your sword. Push him upward and cut: this keeps him from being able to strike at your head.

Push his arms upward with the cut and step through.

This is the same movement as before, but to the other side. You can either step as above, or with your left foot. Which edge to use depends on which side your opponent is attacking from.

If you, with a strike or other technique, bind his sword with strength and he then jerks his sword away and strikes at your head from above, turn your sword with the crossguard in front of your head and cut him from underneath through his arm. With the cut, place the point of the sword on his chest from underneath.

Item wann du im mit ainem haw oder sunst starck uff sin schwert bindest laust er dann sein schwert abschnappen von dem dinen und schlecht dir oben zu dem kopff so verwende din schwert mitt dem gehültz für dein haupt und schnyd im unden durch sin arm und setz im mitt dem schnitt den ort unden an sin brust

As he lifts his sword from the bind to strike from above, simply wind to either side. Step in time with your cut. Make sure that you push your blade forward as you wind so that you get a good cut in the same movement.

As either cut is finished, your point will be on his chest for a thrust.

This can also be said to be a case of nachrayßen.

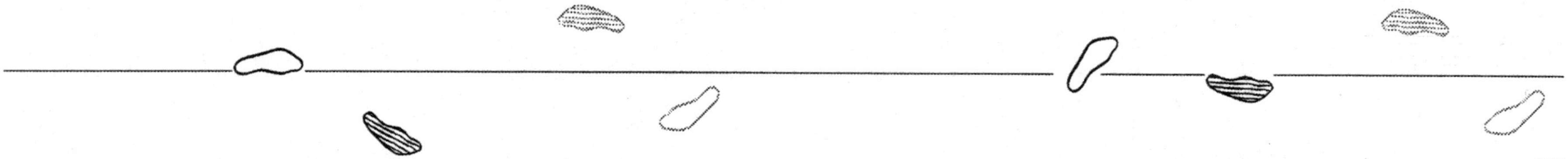

Do the cut like this. If he binds your sword on your left side and strikes around to your right side with a *zwerchaw* or another technique, then leap with the left foot away from the strike to his right side and hit with the true edge over both his arms from above. Practice this on both sides.

Item den schnidt tryb also wenn dir ainer an das schwert bindt gegen diner lincken sytten und schlecht umb vom schwert mitt der zwerche oder sunst dir zuo der rechten sytten so spring mitt dem lincken fuß uß dem haw uff sin rechte syttenn und fall im mitt der langen schniden oben uber bayde arm Das tryb zu baiden syttenn

Binding on your left can mean a strike from the extreme left or simply that he is angling the strike more from that side. As he begins to strike around, step to your left and simply lift and strike with your long edge over his arms with force.

Do not step forward too much; sidestepping is better. If you wish you can pull your rear foot with you in a small step.

Since your sword is placed over his arms his strike will be stopped short. Press his arms down with the cut, or push forward with a cut. You can then cut his throat or face, or push him away.

"Turn the edge and press the hands."

If you can use the lower cut at his arm as you close with him so that the point of your sword points to his right side, then push up with force in the cut. As you push, leap with the left foot to his right side and wind your sword with the long edge up over his arms so that the point of your sword points to his left side. And push his arms away from you.

"Die schnyde wende / zu schlechen druck die hende"

Daß ist wann du im in dein einlauffen kanst mitt dem undern schnitt unden in sin arm daß dein ort gegen siner rechten sytten außgät so truck mitt dem schnitt vast übersich / und mitt dem trucken spring mitt dem lincken fuß uff syn rechte sytten und wende din schwert mitt der langen schniden oben über sein arme das din ort gegen siner lincken sytten uß gee und truck sin arm domitt von dir

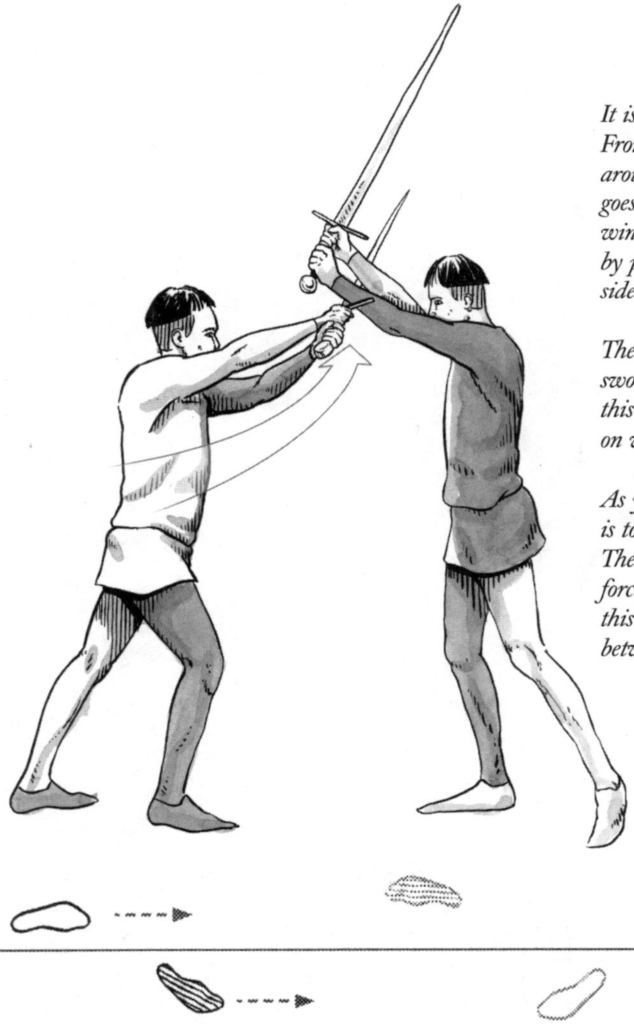

It is possible to begin this technique from several points. From a bind you can move in as he tries to come around. Or you may leap forward into his attack as he goes high. Lastly, it is of great use in the krieg *as he winds. Close with him using a lower cut, push him up by placing your sword on his forearms, and cut to the side that the point is on while you push upward.*

The major reason to step to the left is to get clear of his sword. It is possible to perform the technique without this movement. You can step with either foot depending on whether you step or pass with the rear foot.

As you step, simply turn your sword so that your point is to your right side. This will pivot it around his arms. Then cut down diagonally and toward you with a forceful drawing cut. Or simply push him with a cut; this is less efficient but will increase the distance between you.

Zway Hengen

THE HENGEN (or *hangen*) is a special technique of moving the sword, a bit like *winden,* actually. From *hengen* you can do a lot of different attacks and deflections. The *hengen* takes its name from the fact that you hang the point or the pommel toward the ground in order to gain enough of an angle to threaten your opponent's face with the point. By lowering the pommel, you can position the sword for a steeply angled thrust from below that is very difficult to defend against. From above, do the *hengen* by lifting the pommel and hanging the point down toward his face.

RINGECK ALSO STATES that you can wind, cut, or strike as well from these positions. This means that you must not limit your repertoire to only these simple thrusts, but may in fact do every technique, just as from the *winden.*

IT IS IMPORTANT to distinguish here that Ringeck does not seem to indicate only the specific techniques in his manual when he refers to what you can do from the *hengen* or *winden.* Rather he refers to the three wonders: strike, cut, and thrust. And he seems to imply that they, along with the *hengen* and *winden,* constitute the core of the art of the longsword. This makes sense, since it reduces the components to as few parts as possible.

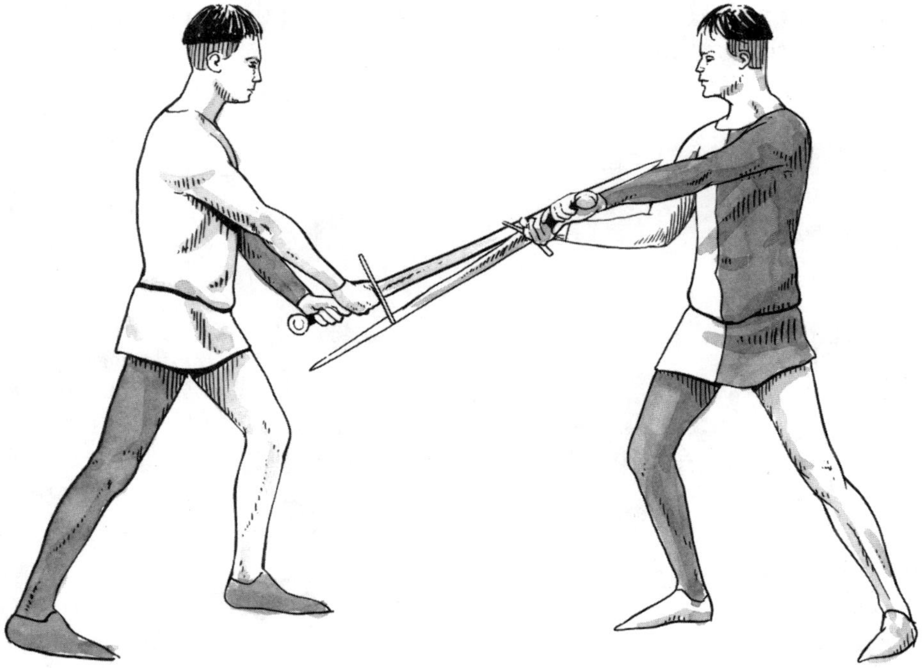

Hengen *means hanging, and in this case it is the pommel or point that hangs downward.*

"There are two hengen, *with one hand from the earth do in all, strikes, thrusts, and guards weak or hard."*

Note: There are two *hengen* on one hand, and from one side from the ground do it like this. If you bind him at the sword with the lower displacement on your left side, let the pommel of your sword hang to the ground and stab him in the face from below. If he then deflects your sword point by lifting it above him, then remain at the sword and lift up also. Let the point of your sword hang toward his face from above. And from both *hengen* you will practice diligently all techniques, strikes, cuts, and thrusts as you feel in the bind if he is weak or strong at the sword.

"Zway hengen werden / uß ainer hand von der erden in allem gefert / hew stich leger waich oder hert"

Mörck eß sind zway hengen uß ainer hand und von ainer sytten von der erden die tryb also wann du im mit dem undern absetzen gegen diner lincken sytten an sin schwert bindest so heng dinen schwert knopff gegen der erden und stich im von unden auff uß dem hengen zu dem gesichte stost er dann mit versetzen dinen ort übersich uff so blyb also am schwert und far mitt im uff und heng im den ort von oben nider zu dem gesicht und in den zwayen hengen solt du alle gefert behentlich tryben hew stich und schnitt dar nach alß du in dem anbinden der schwert enpfindest ober damitt waich oder hört ist

You can hang the pommel from many positions and deflections. It is important that your sword is on top of his and the point is aimed at his face.

Thrust straight up at his face, taking a short step with the lead foot if needed. Thrust short, using the body to drive the point home.

If he parries as below or by lifting the crossguard with the point up, remain on his sword and discern if he is strong or weak.

Without stepping, wind the sword so that your point is at his face and thrust into it or at his chest.

Deflecting with the hengen *is done on the flat of the blade and on the strong part. This leaves the blade free to strike around or thrust; if done with the edge, it will not allow the opponent's weapon to slide off.*

Sprechfenster

THIS IS RINGECK'S version of an extended middle guard, and a transitional position employed when crossing swords with an opponent. It is very versatile, with its only real weakness being that the only attack you can make without much sword and /or body movement is a thrust. Either foot forward works. Remember to keep the sword point aimed at the opponent's face.

THE SPRECHFENSTER IS ideal for feeling out your opponent, since it gives the opportunity for easy defense and blade contact by listening with your sword. It is also a very useful guard for countering the *pflug*, and from it you can easily move close in. But its real benefit stems from the ease with which you can maintain distance while feeling out your opponent. The *sprechfenster* can be assumed before the *zufechten* and then contact is made, or you can assume the *sprechfenster* from an *absetzen,* or a deflection in the *pflug.* However, you must be aware that, as with the other positions that extend the sword point, there is a danger inherent in the position itself, of your opponent's attacking your sword successfully and then closing in.

YOU CAN HOLD the sword even further out than this, but it will limit your options while keeping the opponent further away.

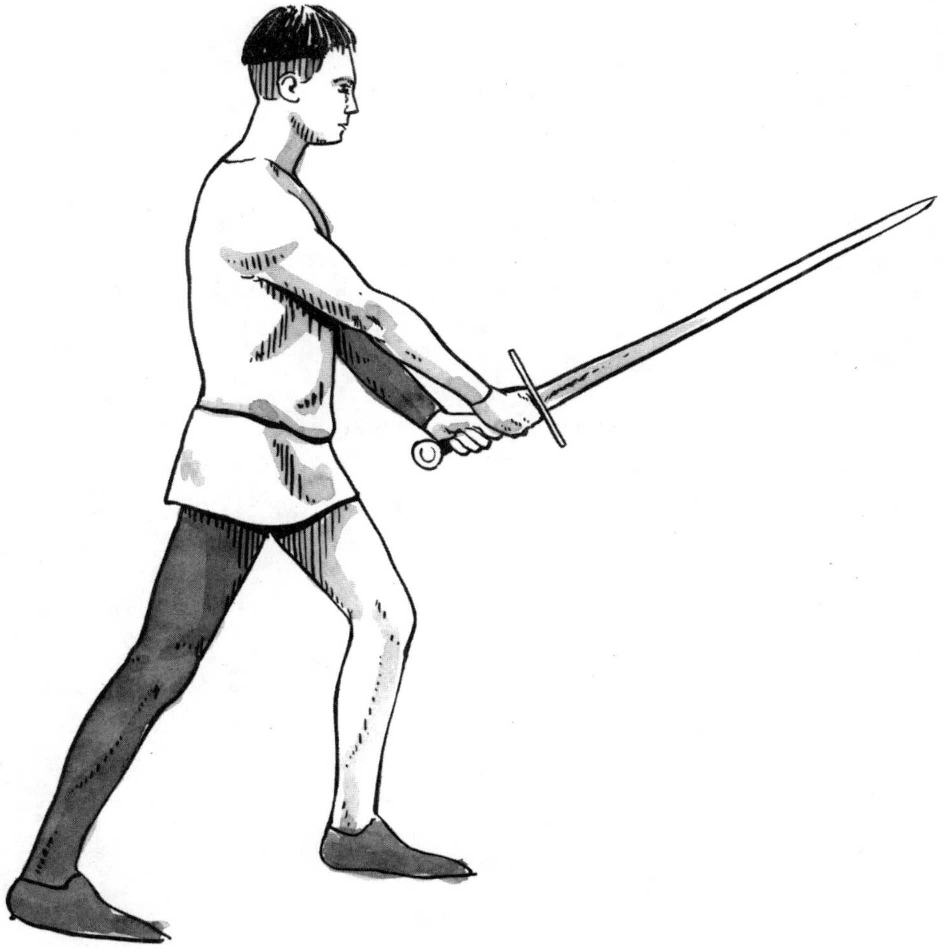

Sprechfenster *means talking window, from the words* sprech *(talk) and* fenster *(window). It is an opening through which you can communicate.*

"Do the sprechfenster *and complacently watch his actions and strike down as he snaps. When he withdraws from you, I honestly tell you, no one defends without danger. If you have understood this, he will not come to blows."*

Note: It is called the "speaking window" (*sprechfenster*). If he binds your sword with a strike or a parry, then remain strong at the sword with extended arms using the long edge. Keep the point of the sword at his face and calmly note what he intends against you.

Note: If he strikes around with an *oberhaw* to your other side, then follow and bind strongly with the long edge toward his head. Or if he strikes around with a *zwerch*, then hit on his arms with the high cut. Or if he yanks the sword to himself and wishes to thrust at your lower openings, then travel after him at the sword and thrust high.

Note: If he neither wishes to withdraw nor strike around, then you will work with the sword using *duplieren* and other techniques as you find out if he is weak or strong at the sword.

"Sprechfenster mach stand frylich besiche sin sach schlage in das er schnappe wer sich vor dir zühet abe ich sag dir für ware / sich schützt kain man on fare haustu recht vernommen zu schlage mag er klain kummen"

Mörck das haisst das sprechfenster wenn er dir mitt hewen oder versetzen an das schwert bindet so belib starck auß gerechtenn armen mit der langen schniden am schwert mitt dem ort im vor dem gesichte und stand freylich und besich sin sach was er gegen dir tryben wöll

Item schlecht er vom schwert umb mitt ainem oberhaw dir zu der anderen sytten so bind mitt storck sinem haw nach mitt der langen schniden im oben ein zu dem kopff oder schlecht er umb mitt der zwer so fall im mitt dem obern schnitt in die arme Oder zuckt er sin schwert an sich und will dir unnden zu stechen so raise im nach an dem schwert und setz im oben an

Item oder will er sich vom schwert nicht abziechen noch umbschlagen so arbait du am schwert mitt dem duplieren und sunst mit anderen stucken darnach alß du empfindest die waich und die hört am schwert

Lang Ort

The lang ort or long/extended point, is a measure that Ringeck suggests you use if you find that you are too close to the opponent and wish to create space between you. It can also be used to keep an opponent at a distance, but there is a distinct risk that he can mount a successful attack on your blade to move the point aside and create room for an attack. Be aware that your opponent must attack your sword first–there is no other way he can get to you. Therefore, if you are in the *lang ort*, any attack against you will be directed at the blade.

Be prepared to *durchwechseln* with the sword and attack with a thrust.

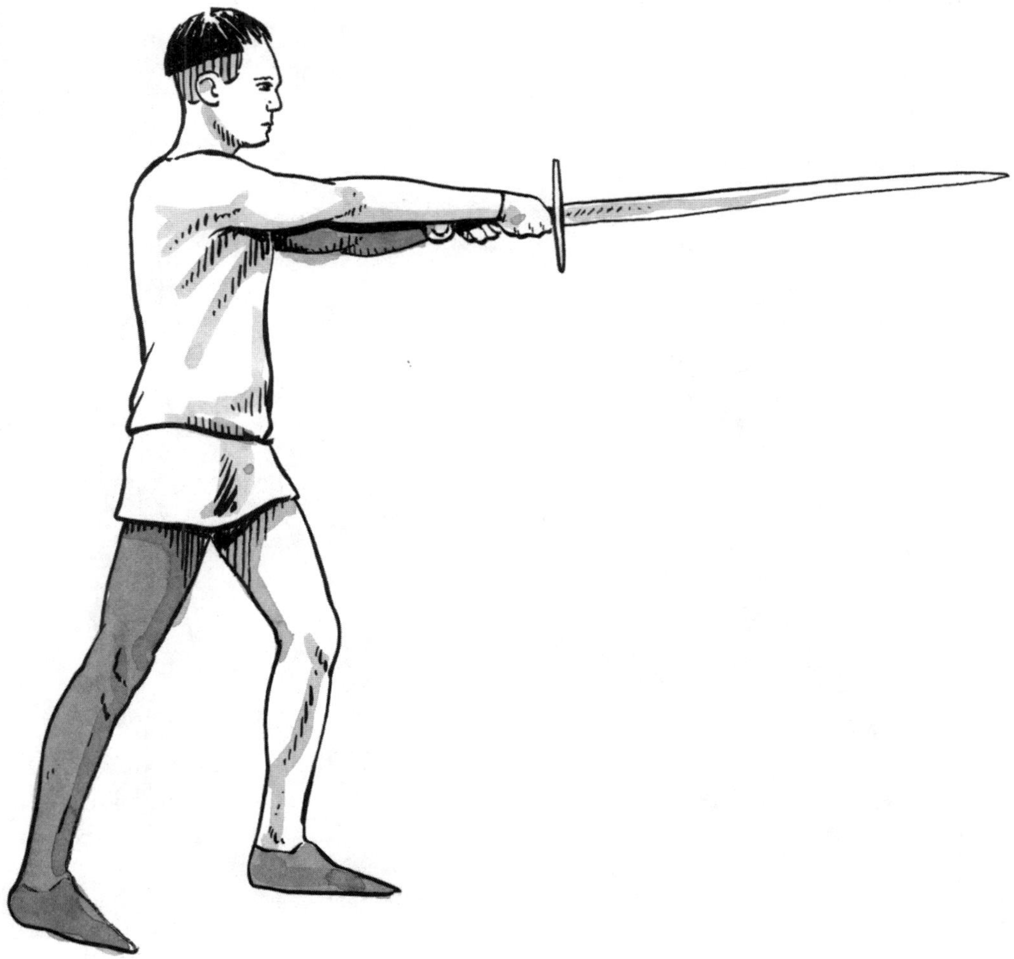

Lang ort *means long point.*

"Here note why it is called the lang ort."

Note: When you come close to him, place your left foot forward and extend your arms with the sword point at his breast or his face. If he then strikes from above at your head, turn your sword against the strike and thrust at his face. Or if he strikes from above or below at your sword and wishes to knock your blade aside, then change through and thrust at him in the opening on the other side. But if he hits your sword with force, let it snap around and hit him thus at once on the head. If he moves in, wrestle or cut him. Make sure you do not fail.

"Hie nach mörck waß do haysset der lang ort"

Mörck wenn du mitt dem zufechten zu nahent an in kümst so steze dinen lincken fuß vor und halt im den ort auß gerechten armen lang gegen dem gesicht oder der brust hawt er denn dir oben nider zu dem kopffe so wind mitt dem schwert gegen sinen haw und stich im zu dem gesicht oder hawt er von oben nider oder von unden auff dir zu dem schwert und will dir den ort wegk schlachen so wechsel durch und stich im zu der anderen sytten zu der blöß oder trifft er dir mitt dem haw das schwert mitt störcke so lauß din schwert umb schnappen so triffest du in dem kopff laufft er dir ein so tryb die ringen oder den schnitt Lug das es dir nitt fel

You can use the lang ort *if you are too close to create space, or as a way to approach your opponent. If you feel you are too close, put your left leg foremost and extend your arms as in a thrust aimed at your opponent's chest.*

If he strikes at you, wind your sword to meet his strike. You can wind in many ways–this is one possibility. You can also meet his sword as you would from the vom tag *with a* zornhaw *counterstrike.*

Thrust at once at his face. Take a small step forward with your lead foot and make a quick thrust. Here your arms are crossed; if you meet in the zornhaw *they will be uncrossed.*

As he strikes at you, simply change through to the side from which he began his strike and attack him there high. Changing through is relatively easy since he is forced to attack your sword point and thus will strike well away from your body, giving you space to move. Drop the point under his attack as you move in.

Thrust at him at once in the exposed opening on the side he struck from.

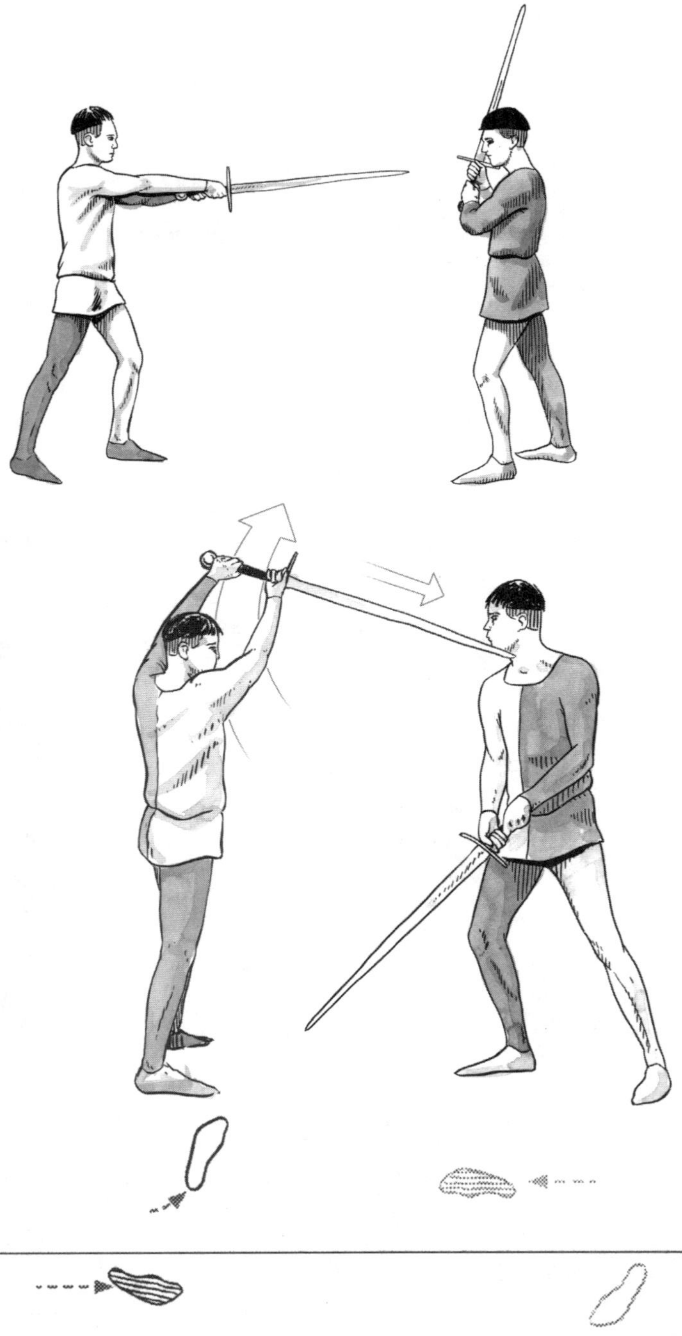

If he strikes your sword with strength, go with the force and do not resist. Instead let your sword begin to swing around, going with the force and gaining speed.

Swing around and strike him in the head, with or without taking a step. You can also hit his hands or lower arms.

Beschließung der Zedel

THIS SUMMARY SEEMS to have been included after the compilation of the original rhyme by Liechtenauer. And yet it retains something of the characteristics of a rhyme. In essence it does not introduce anything new, but only recapitulates earlier statements. Probably it had been included in the wrong place in the existing version due to a clerical error in the copying process. The paragraph resembles most of all the summary you would expect to find at the beginning of the text.

IT IS EVIDENT in the next paragraph that the editor or writer of the text was copying something that was not in the right order, since the next one again deals with the *hengen* and *winden*, which had already been covered in the material.

"The teaching summarized."

In the following the teaching is summarized. You will always be ready and trained in the art of fencing so that you can use all counters against your opponent's techniques. From each counter you will work with the three wonders (strikes, thrusts, and cuts). You must know *hengen* with the sword, and from the *hengen* the eight *winden*. And from the *winden* you must also know how to use the three moves (strike, cut, and thrust).

"Der beschliessung der zedel"

Das ist ain lere do die zedel kurtz inen begriffen ist die lert also daß du in der kunst gar wol bericht und geüpbt solt sin gegen den do du mitt fichtest daß du dine brüch gegen sinen stucken also wisset zü tryben daß du uß ainem yeden bruch besunder magst gearbaitten mitt dreyen wunder auch solt do wol hengen am schwert und auß den hengen solt du bringen acht winden und die winden solt du also betrachten daß du der aines selbs dritt wissest zu tryben

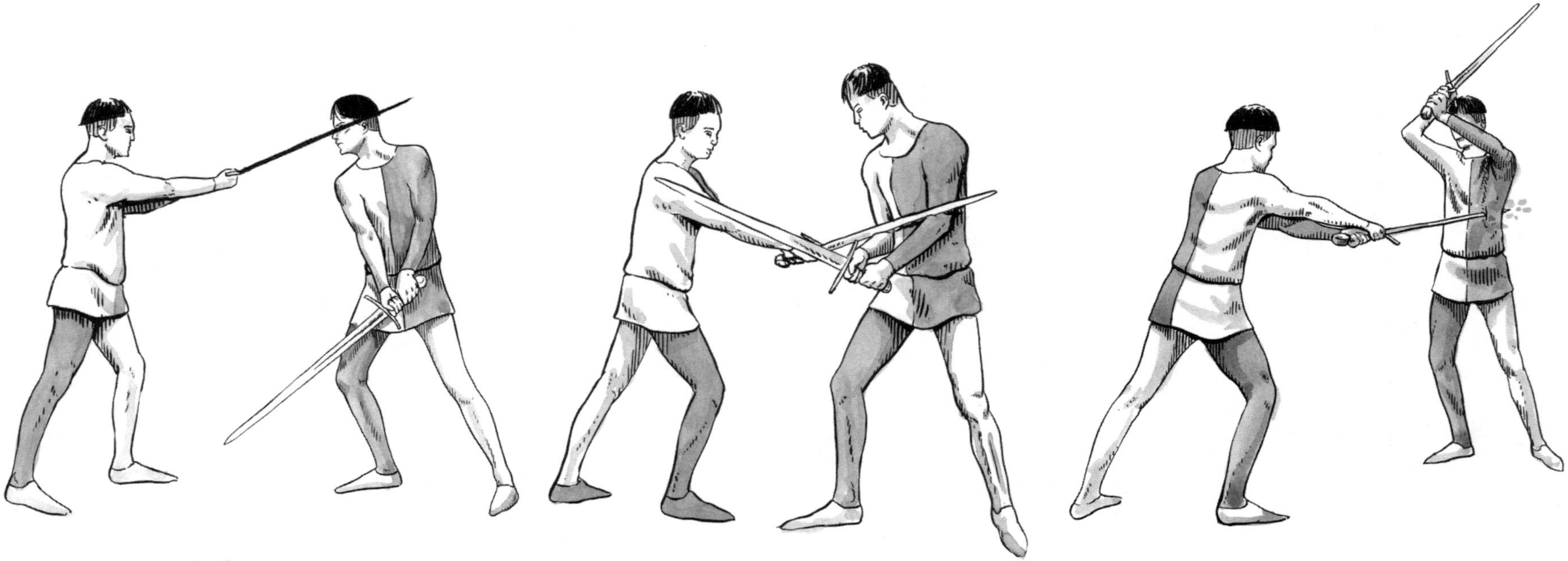

Strike *Cut* *Thrust*

"Here note how you will try the hengen *and the* winden."

Note: There are four binds at the sword, two upper and two lower. For each bind you will master two *winden.* Do it like this. If he binds at your upper left side, wind the short edge at his sword and lift up your arms. Let the point of the sword hang and thrust at his face. If he deflects the thrust, let the sword point hang down on the sword and wind on your right side. These are the two *winden* on one side of the sword.

Also, if he binds up at your right side, wind the long edge to your right-hand side against his sword. Lift your arms and let the sword point hang and thrust at his face. If he parries the thrust with strength, let the sword point hang inward, wind on your left side, and thrust at him. These are the four *winden* from the two upper binds on the left- and right-hand side.

"Hie mörck wie du hengen und die winden treyben solt"

Das vernym also es sind vier anbinden der schwert zway obere und zway undere nun solt du uß yedem anbinden der schwert besunder trybenn zway winden das vernym also bindt er dir oben an gegen diner lincken sytten so winde die kurtz schnide an sin schwert und far wol uff mitt den armen und heng im den ort oben ein und stich im zu dem gesicht. versetzter den stich sol auß den ort am schwert oben ein hangen und wind uff dein rechte sytten daß sind zway winden an ainer sytten des schwerts

Item oder bind er dir oben an gen diner rechten sytten so wind och gegen diner rechten sytten die langen aschnid an sin schwert und far wol uff mitt den armen und heng im den ort oben ein und stich im zu dem gesicht versetzt er den stich mitt störck so lauß im den ort am schwert oben ein hangen und wind uff din lincke sytten und stich daß sind vier winden uß den obern zwayen anbinden von der lincken und von der rechten sytten

Also you will know four *winden* from the two lower binds with all techniques, making eight. And remember that you can use strikes, thrusts, and cuts (the three wonders) from each *winden* so that you can do a total of 24 techniques from the *winden*. You will learn eight *winden* on both sides, so that with each wonder you can try to see if he is weak or strong at the sword. When you understand this, then use all techniques that belong to the *winden*. If you do not do this, you will be defeated at every *winden*.

Item nun solt du wissen das du uß den undern zwayen anbinden auch solt tryben vier winden mitt allen gefertenn alß uß den obern also worden der winden oben und unden acht und gedenck das du uß yedem winden besonnder triben solt ainen haw oder schnitt und ain stich und das haissen die drew wunder darumb daß man sy zu vierunndzwintzig malen uß den acht winden tryben sollund mag und die acht winden solt du wol lernen tryben von baiden sytten also daß du mitt yedem wunder brieffest sin gefört nicht mer dann öb er damitt waich oder hert am schwert sy unnd wenn du die zway ding enpfundenn haust so tryb die stuck gerecht die zu dem winden gehören wo du das nicht tust so wirst du by allen winden geschlagen

Nebenhut

NEBENHUT (from *neben*, meaning lower, and *hut*, meaning guard or position) is not considered one of the primary guards, but Ringeck says that it is good to fence from it. It can only be done with any degree of effect from your strong side. The reason for this is that your rear hand will lose its grip on the pommel if you try to pull the sword back on the off-hand side while keeping the long edge forward. You do not hide the sword or make it hard in any way to guess where it is going to come from, as has been suggested. What this position does is offer an imagined opening to your opponent–it is, in a sense, a trap. It is also a difficult guard to fence well from.

TO COMPLICATE THINGS further, Ringeck states that you should only fence from the left side if you are right-handed. Now this goes against what has just been said. But Ringeck does have a point, if you consider the specific counters that he proposes. If you hold the sword on your left side against a right-handed opponent, it is also very easy to counterstrike because you will be able to lift his strike from underneath with ease. If we compare this with the later work of Joachim Meyer in 1600, we find a guard called the *wechselhut*, the position that you end up in on the left side after a *zornhaw* from the right. There is no *nebenhut* on the left side, but on that side we have a *wechselhut.* A possible interpretation of this guard in Ringeck is *nebenhut* on the right side and a *wechsel* on the left.

IT MAY SEEM odd that the techniques should primarily be performed from the left side rather then the right, since Ringeck in his introduction emphasizes that right-handed fencers should fence from the right side. The logic behind this is that the *nebenhut* techniques would (primarily) be used after a descending strike from the right side, which would place the sword on the left side for a right-handed swordsman. It also has to do with the cutting angle if your opponent is right-handed; if he is left-handed you must work from your right side to achieve the same result. It is important to remember that the techniques from *nebenhut* seem to be counters rather than attacks. This reinforces the assumption that the *nebenhut* techniques are primarily intended as follow-ups or continuations of other attacks that may end with the sword low, or in a position where it is very easy to lower the sword in a continuous movement. It is also a position in which the opponent can be convinced that you are vulnerable. However, it has nothing to do with either hiding the sword or obscuring from where an attack may come. Remember that you can actually use the principles presented regardless of the side you fence from or, rather, regardless of the side you end up on. You must practice with a sword in hand.

"Here note how to strike and fence from the nebenhutten.*"*

Know that from the strikes (that is, from the lower guards) it is good to fence. Even if they are not named in the teaching, all techniques derive from the teaching that you use to fence. And the strikes will be done from the left side, since they are not as effective from the right side.

"Hie mörck uß den nebenhutten daß den strychen zu fechten"

Wiß das auß den streychen gut fechten ist wie wol sy in der zedel nicht benampt sin so kommen doch die stuck ußß der zedel die man daruß vichtet und die streychen soll man treybenn von der lincken sytten wann von der rechten sind sy nitt also gewiß alß von der lincken

On the left side a wechselhut.

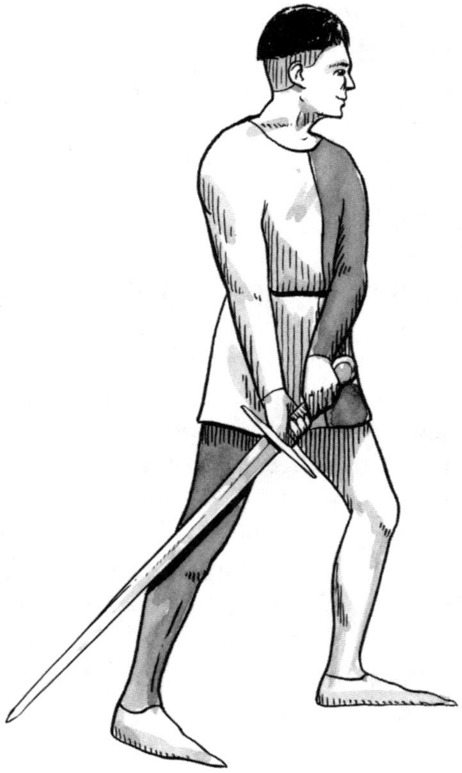

On the right side a nebenhut.

The position of the nebenhut *on the left side in Ringeck corresponds to* wechsel *in Joachim Meyer's work of 1600. Lead with the right leg and hold the sword aimed slightly forward, as if you just completed a* zornhaw*. On the left side, you will lead with the short edge, and on the right side the long edge will be forward.*

It is important that the edge is pointing toward your opponent while in this guard; otherwise, precious time will be lost when making the strike in turning the sword around.

Nebenhut *is simple and effective. The most common fault is pushing the rear elbow too far, past the back, thus locking the arm in position. Be sure to just move the sword to your rear, and when your rearmost arm can go no further, leave it there.*

Keep the long edge facing forward and aimed at your opponent.
Do not pull the sword so far back that your elbow goes back past your waist; this will slow you down.

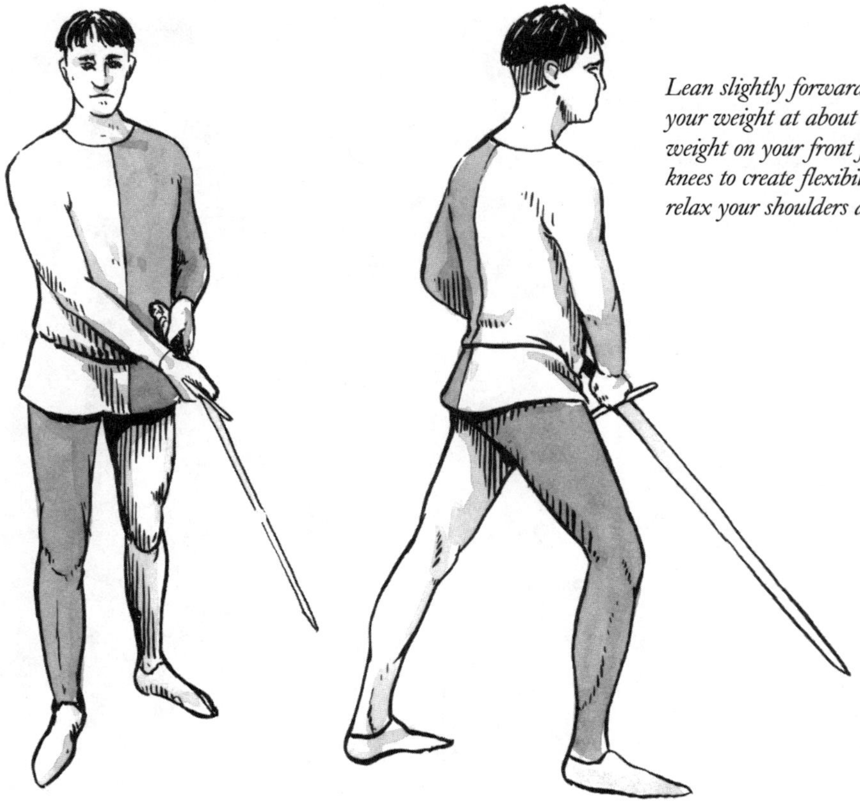

Lean slightly forward, but try to keep your weight at about 60/40 with more weight on your front foot. Bend your knees to create flexibility and be sure to relax your shoulders and arms.

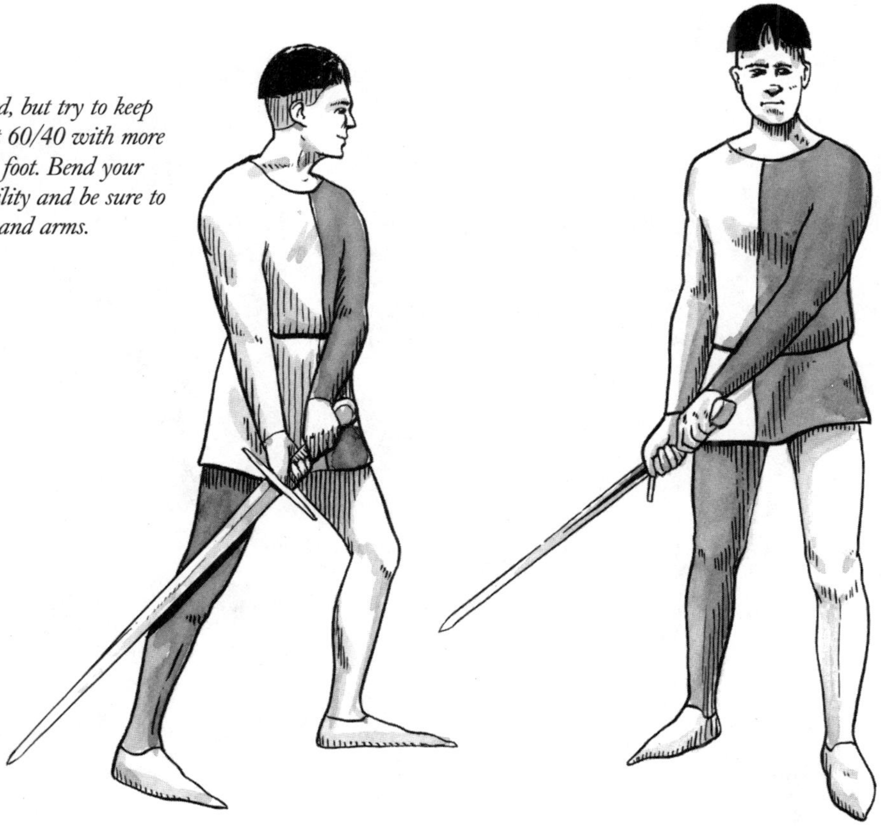

If you stand in the lower guard on your left side and he strikes with an *oberhaw,* then strike from below at his sword using the short edge. If he holds strongly against this and is not too high with his hands, *duplieren* between the man and his sword using the short edge to the left side of his neck.

Item wann du ligst in der nebenhut uff diner lincken sytten unnd ainer hawet uff dich von oben nider so streych von unden uff vast in sin schwert mitt der kurtzen schniden helt er starck wider und ist nicht ze hoch mitt den henden so duplier zwischen dem man und sinem schwert ein mitt der kurtzen schniden zuo sinem lincken halß

As he strikes, lift your sword sharply to knock his blade to the side. Do this at a slight angle, as if aiming your sword to end up at your right shoulder. You will ideally impact on the flat of his blade rather than the edge; otherwise you may be struck as your sword glances off his.

Duplieren *with the back edge to his neck. If you do it like so, he will not be able to get his sword in between or sidestep. Or you can push your sword forward and cut his throat as you step to his outside.*

This attack is to the right side, which as a matter of fact is easier to hit, but you can also push your sword against his neck and cut by either drawing or pushing the blade.

If you strike at his sword as before and he is weak at the sword and his hands are low, strike him at once at the upper openings using the long edge. Or if it seems that he is striking with force with his sword on yours from above, move your pommel over his sword, keep your hands on it, and aim the point of your sword back to your left side, and let the sword snap around and strike him in the head using the short edge.

Item wann du uff streychst alß vor an sinem schwert ist er dann waich am schwert und nider mitt den henden so haw im zu hand mitt der langen schniden oben zu der blöß Oder fölt er dir starck mitt dem schwert uff daß din so far im bald mitt dem knopf über sin schwert und bleyb mitt den henden dar uff und laß den ort hindersich zu diner lincken sytten unnd schnapp im mitt der kurtzen schniden zu dem haupt

This is a snapping motion where you strike using the wrist and elbow rather than the whole body. It is not a strong strike, but surprising and very quick. It can be done from either side.

Since your opponent is pressing on your sword, he will lose initiative and his balance when you remove it.

Snap the sword around, at the same time taking a small step forward and to the side with your lead foot. Strike his head or neck. You can also step farther to the side and strike him from the side in the neck or head.

If you strike at his sword and he lifts the sword high and winds, take a step back and strike him with arms extended to the right.

Item wenn du im streychst an sin schwert fört er hoch uff und windt so schlach in in rechten sytten mitt gesträckten armen und tritt damitt zu rucke

As he strikes, you strike at his sword from below. This works on both sides, and you can move in several ways depending on where you wish to go next.

Instead of binding hard, your opponent goes with the strike and begins to wind high. He may also go into hengen, *in which case the sword will be more vertical.*

As he winds to the side, step to the side and aim a horizontal strike at his arms or shoulder.

If you strike at his sword and he lifts the sword high and winds, then strengthen your bind while you wind to the long edge. If he then strikes around with a *zwerchaw*, strike at his left side while stepping back.

Item wann du im uff streychst an sin schwert fört er hoch uff und windt so sterck mitt der langen schniden schlecht er dann umb mitt der zwerch so schlach in in die lincken sytten mitt ainem abtritt

After you have struck his sword, he winds. You can then either wind yourself, or choose to strengthen the bind without winding; this yields different options. An advantage of winding is that you can step out to the side and be in a good position to strike at him.

Since you are also winding, you will turn his point to the side, forcing him to change his attack. He then tries a zwerch.

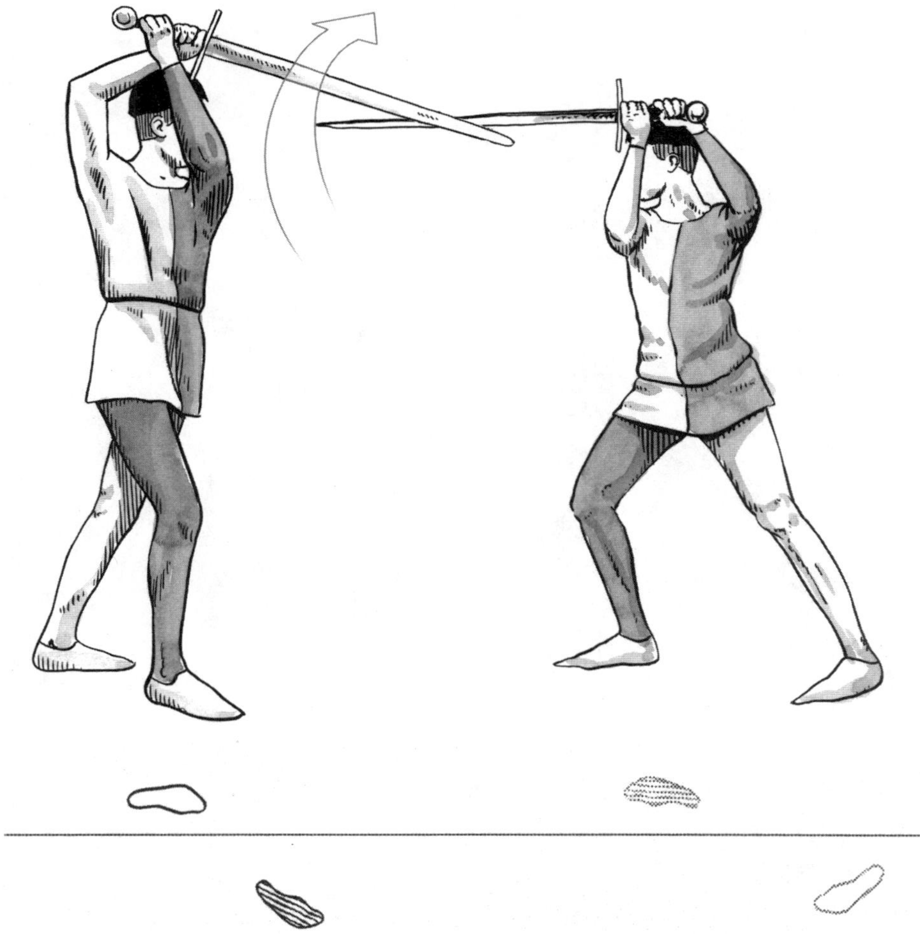

As he aims a zwerch *to your side, step back and strike horizontally at his arm or shoulder. It is also possible to change through and thrust to his open side.*

If you try to strike at him and he holds the sword diagonally (*zwerch*) in front of him with his arms high and wishes to fall on your sword, strike him from underneath at the sword and strike his arm or thrust at his breast. (Note that *zwerch* also means diagonally and, in this case, I think it likely that it refers to any position when an opponent holds the sword in such a diagonal position. I find it unlikely that this interpretation refers to the *zwerch* strike.)

Item wann du tribst die streychen zu dem mann und helt er dann sin schwert zwerchs vor im und ist hoch mitt den armen und will dir uff din schwert fallen so streych im unden an syn schwert und schlach in uff den armen oder stoß in an die brust

Strike upward with the short edge, either at his sword to get it out of the way and position your sword for a thrust, or to strike his hands from below.

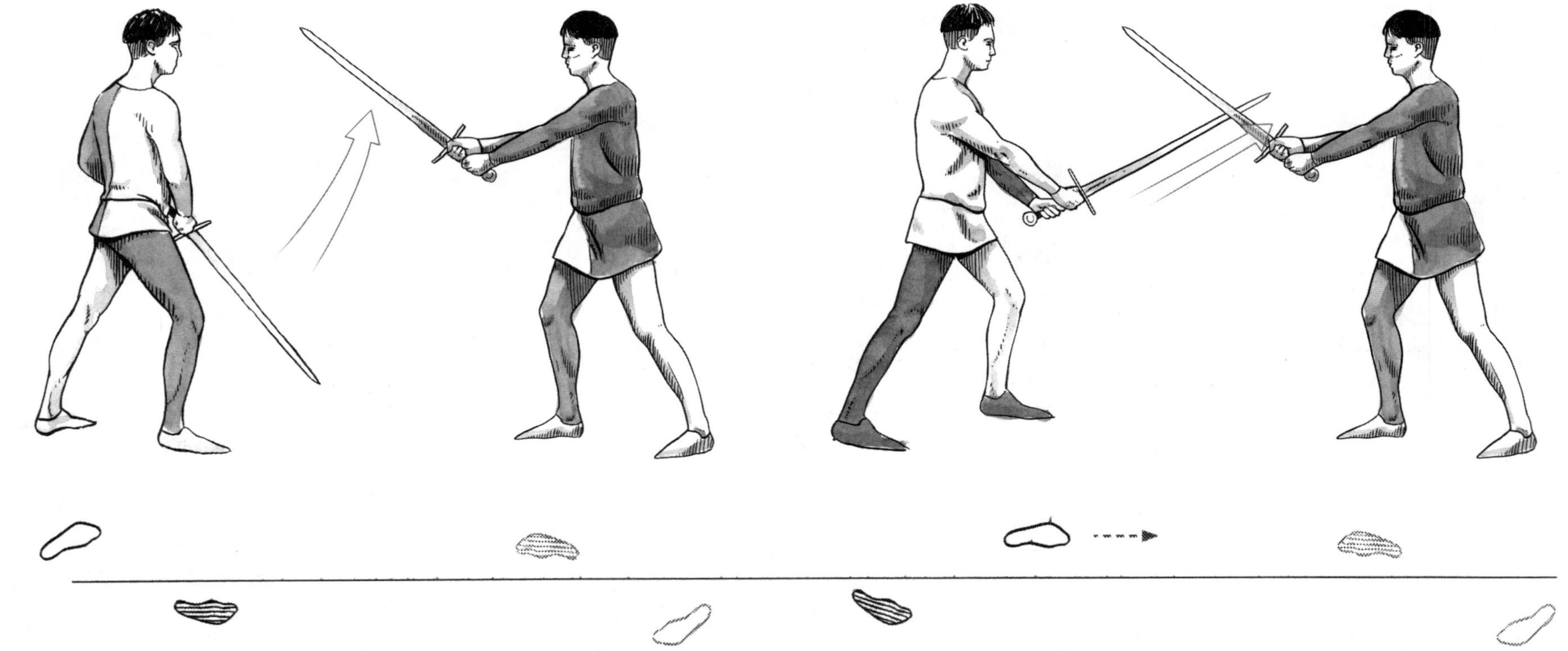

Strike his hands by simply lifting your sword. Step out to the right as you strike with the short edge.

At once after the strike, take a step forward with the lead foot and thrust at his stomach or chest, depending on where the opening is.

If he keeps his hands low and wishes to come down on you, strike through to the other side and thrust him in the breast: this is to change through (*durchwechseln*).

Item ist er nider mitt den henden und will dir uff fallen so streych durch uff die anderen sytten und stoß in zu der brust / daß ist durch gewechselt

Strike upward with the short edge and step forward to either side with your lead foot. Wind in the strike and thrust at his face. You can do the same after you have struck a descending strike to the side.

You can also strike to the other side more horizontally and thus thrust at in the chest with the sword angled inward and horizontally. If you miss his sword, then wind and thrust.

If you strike through, come down with the long edge on his sword and wind to your left so that your thumb is down. Place the strong of the long edge on the right side of his neck and leap with the right foot behind his left and jerk him over it with the move.

If you change through from the strike and end up at the upper part of his sword on the other side, then you can do all the techniques as before with the *zecken* and every other thing.

Item wann du durch streychst so fall im mitt der langen schniden uff sin schwert unnd wind uff din lincke sytten das din daum unden kompt und far im mitt der langen schniden an sinen rechten halß mitt der störck und spring mitt dem rechten fuß hinder siner lincken unnd ruck in mitt dem schritt daruber

Item wann du uß dem streychen durch gewechselt unnd hompst zu der anderu sytten oben uff sin schwert so magst du die stuck glychals wol tryben als vor uff gener sytten mitt zeckrüren und mitt allen dingen.

Do not place your sword on his–slam it down. This will knock him off balance and give you time to strike.

Step to the right and guide your point through to your right side. Let the sword continue with the wind so that your long edge is on his sword. You can also step with your left foot, or the technique can be done without moving in first.

Step forward with your right foot behind his left and slam the strong of the long edge into his throat; do it as fast as you can and, in the same move, heave him over during the cut.

Absetzen

ABSETZEN MEANS "SETTING ASIDE." This is not a position, but a principle with which you can defend and counter well. It is executed from the *pflug* and can be used against both thrusting and striking attacks. Ringeck emphasizes that we do not need to go anywhere in order to fight from the *pflug*, or any other position for that matter. *Absetzen* is a key movement for fencers to understand. They must be able to do it well from the *pflug* because it is the key exit from this position into the counterattacks. In parrying, the flat must be used along with a turning motion to the side in time with a step forward. In the movement, the parry and the attack blend into a single action, not two. From the *absetzen* openings exist into every other move, usually by going into a bind and then winding or changing through to create a new opening.

Absetzen *means to set something aside, for example a sword thrust. To change through with the point, use the pommel; do not move the arms. Use the pommel and keep the front hand still, and you will be able to change openings very quickly. Note that Ringeck states that this works on either side.*

"Note how to fence from the absetzen."

If you are fencing with someone and come close to him, assume the guard of the *pflug*. Do this quickly and with *winden* from one side to the other. Aim the point of your sword where you please. From this you can parry; that is how close it is. And you can get stronger with the long edge from this position and perform all the aforementioned techniques. You can also deflect strikes and thrusts and break them with the *winden* and seek openings with the point of the sword.

"Mörck ain zu fechten uß dem absätzen"

Item wann du mitt ainem fichtest und wenn du nachend zu im kompst so kom in den pflug und tryb den behentlich mitt winden von ainer sytten zu der andern und das din ort stets dir belib und uß dem magst du tryben daß versetzen daß ist die nech und zu dem magstu stercken mitt der langen schniden und daruß tryben alle vorgenampte stuck och magstu hew und stich absetzen und die schlechtlich brechen mitt winden und mitt dem ort deß blöß süchen

Schrankhut

This is not exactly a guard in Ringeck's system; rather it is a place that you arrive at from, for example, a strike from below or after a *krumphaw*. And because it is a position that does occur, it is important to know what to do with it. It is not a position in which to await an attack; it is too weak. Hans Talhoffer says in his fechtbuch of 1467 that the left side is weak. This statement is, however, not repeated in any other manuals that actually show only the left side. Both Jacob Sutor and Joachim Meyer indicate the use of the guard on both sides. But remember that while Ringeck says that you should only fence from the four guards, the other positions and possibilities that he mentions are to be employed as necessary. The *schrankhut* can be done on both sides–note that if you lead with the left leg your arms will be crossed, which produces a weaker grip, so the right side is preferable if you have a choice. From this position, it is easy to step in with a cut from below at the opponent's arms or hands. But understand also that you are somewhat exposed to strikes from above. Since you cannot parry with one of the *versatzungen*, a rising counterstrike is a good option. This is also one of the positions that you will never end up in if wearing armor.

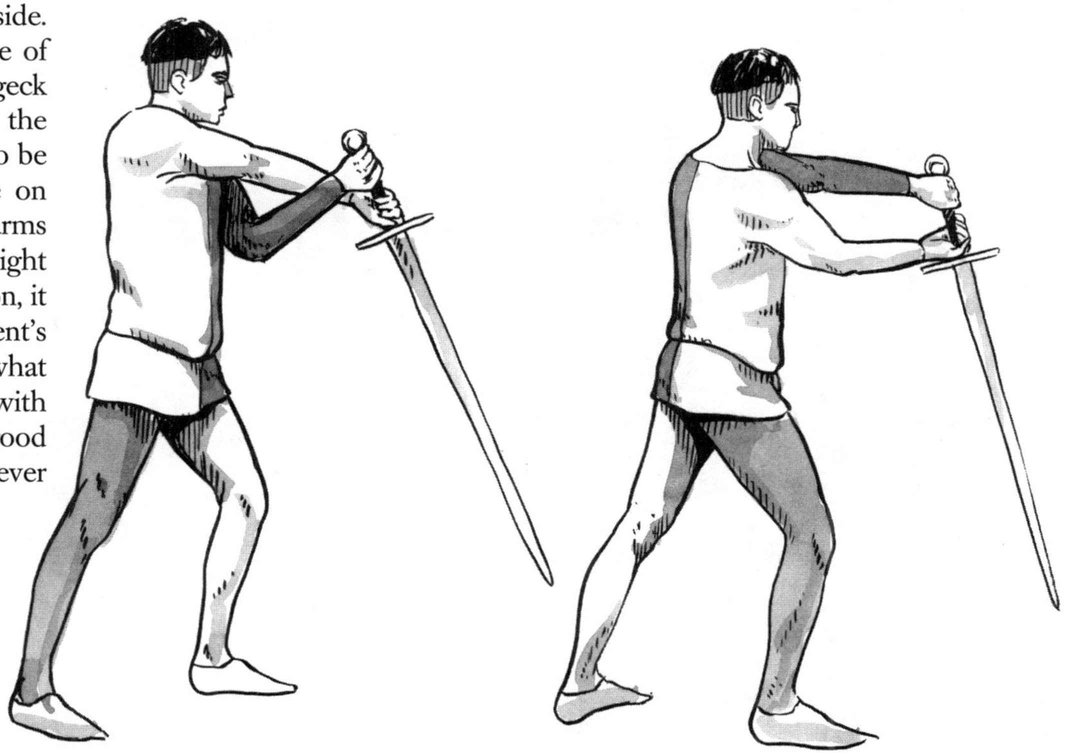

Schrankhut *means fence/railing guard. A* schrank *is a railing or barrier with vertical posts and this is what the position resembles.*

"Do the schrankhut *like so."*

When you fence with someone and come close to him, stand with your left foot forward and put the sword point toward the ground on your right side so that the long edge is on top. And on the left side the short edge is below, and the (right) foot forward.

"Die schranckhutt die mach also"

Item wann du mitt ainem fichtest und nahent zu im kompst so stand mitt dem lincken fuß für und leg das schwert mitt dem ort uff die erde zu diner rechten sytten und das die lang schnid oben sy und von der lincken sytten die kurtz schnid unden / und der fuß vor stee

On the left side.

Do not bunch up your shoulders. It may feel a bit awkward at first, but if you relax you will find that there are many applications possible from here.

Try experimenting with changing how high the crossguard is and the angle of the blade. Different possibilities arise as these change.

On the right side

"Do the technique like this from the schrankhut."

If someone strikes at you high or low, then step toward him and strike a crooked strike (*krumphaw*) at his opening.

Or strike a *krumphaw* at his blade; as soon as the swords meet, strike at once to his next opening with the short edge.

Or turn (*verkerer*) your sword and thrust at his face. If he then binds, hold strongly against it with the long edge. You may also try all techniques mentioned earlier with the strike.

"Die stuck treyb uß der schrankhut also"

Item hawet ainer dir oben zu oder von unden uff so haw du im krump ein zu der bloß mitt ainem ußtrytt

Item oder haw im krump zur flechen und alß bald es klützt so such die nach mitt der kurtzen schnyden

Item oder tryb den verkerer mitt dem ort zu sinem gesicht und wenn er dir anbindt so sterck mitt der langen schniden und magst alle stuck tryben die vor genampt sind in den strychen

You can strike from below in a rising strike, or simply swing the sword around and hit him with an ordinary krumphaw. *Your hands are already in position: just swing the point around over his sword as you step to the right.*

A simple continuation from the krumphaw: *uncoil your hands as you strike around at his head.*

To strike his sword, simply swing upward as he strikes with an oberhaw. *Keep a small angle between the blades; otherwise they might skid off and you will get struck. Do a follow-up with a thrust to his lower opening or wind and strike around.*

As the swords meet, pull the pommel down to bring the point up and thrust under his sword at his face. Or wind to place the point in his face if you are on top of his blade.

Das Redel

THE REDEL (WHEEL) is a way of approaching your opponent. Moving the tip of the sword in a circular motion makes it more difficult for your opponent to ascertain where you are going to strike or thrust. You can also hold the blade more horizontally. The movement can be done as a small circle in front of you or in a larger one that goes above the head.

IT IS ALSO possible to interpret it as a strike from the *nebenhut* proper toward your left side, but reading it as circling makes more sense, since its distinctive feature is that you circle the sword tip as you approach. (And the text does not say strike.)

MOVE THE TIP of the sword in a circle as you approach by working the pommel and grip as a lever so that a small movement of your hand can move the tip a lot. Also circle with your arms slightly to add to the movement. From the *redel* (or *rad*), you are in a position to change through if your opponent tries to bind your sword. Simply drop the point and step diagonally forward and thrust in one movement. Aiming high will more often reward you with a hit.

When you fence with someone, extend your arm, hold your thumb on top of the sword, and move the point of the sword like a wheel from below on your left side in front of you. In this manner approach him. From this you can change through or bind on both sides, and when you have made the bind, you can try all techniques that seem good to you just as before.

Item wann du mitt ainem vechtest so streck din arm lanck von dir unnd daß din daum oben blyb uff dem schwert und wend das schwert vor dir mitt dem ort umb glych alß ain rad von unden uff din lincken sytten behentlich und gang domitt zu dem mann und daruß so magstu durchwechseln uff welche sytten du wilt oder anbinden unnd wenn du angebunden haust so magstu tryben waß stuck du wilt daß dich am besten tunckt alß vor

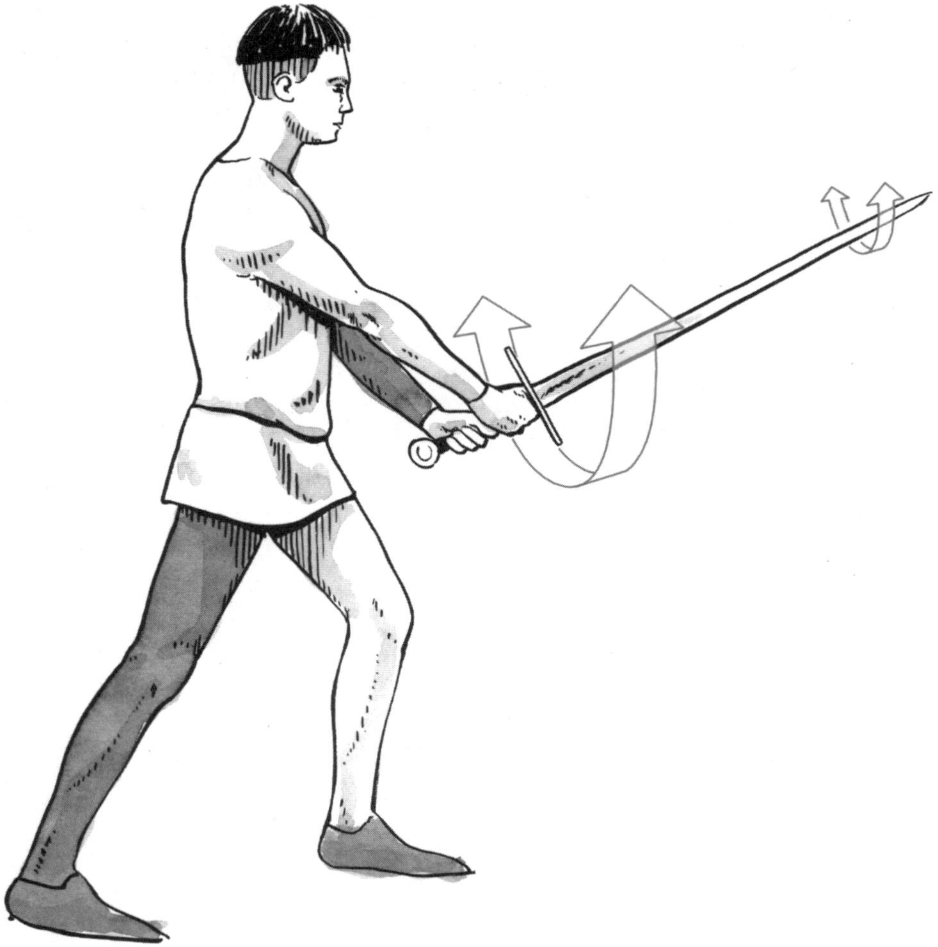

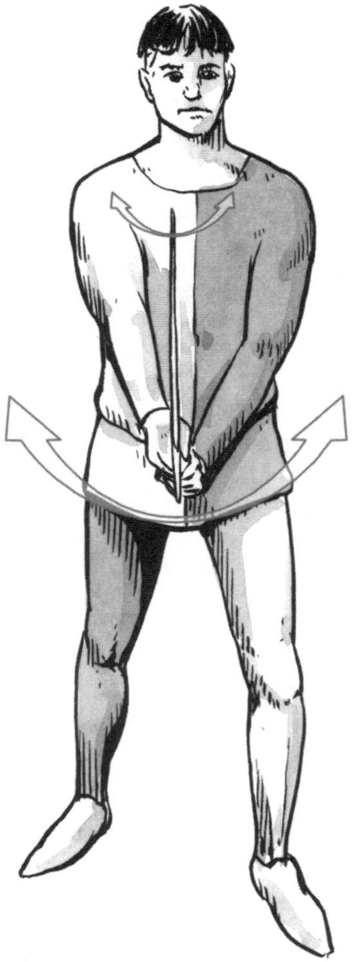

Bruch

The counters, or *bruch* (breaks), are set apart from Ringeck's other comments on techniques. They seem somewhat out of place, and perhaps we should regard them as his very own addition of "Master Ringeck's patent solutions for desperate situations."

They are sometimes useful, but not vital. Their main value lies in that perhaps they are the only unique addition of his own, and provide insight into understanding his style of fencing.

"How to break the zwerch.*"*

If you stand in the guard *vom tag* and someone strikes with a *zwerchaw* at you, then at once strike with a *zornhaw* with the strong at his sword and seek the opening with the sword point. If he then tries to strike around with a *zwerch* at the other side, preempt him with a *zwerch* under his sword at his neck. Or cut at his arm with the long edge as he tries to strike around.

"Also brich die zwerck"

Item wann du stanst in vorhut vom tag und ainer uff dich hawet mitt der schwerch so haw den zornhaw glych mit im ein starck uff sin schwert und such die blöß mitt dem ort und will er denn um sich schlachen mitt der zwerch zu der andern sytten so kom du vor mitt der zwerch under syn schwert zu sinem halß oder schnyd in mitt der langen schnyden in die arm wenn er um schlöcht

As he strikes, counterstrike at his sword. This will push his sword down to his outside. Use the pommel to place the point of your sword so that you can thrust at his face or body.

As he tries to strike around, you can easily reach his arms or his neck with a horizontal strike. Or you can step forward and make an upward cut into his forearms.

"A counter to the counter."

If you strike with a *zwerchaw* and he wishes to preempt you with a *zwerch* under your sword toward your neck, then come down forcefully with the long edge on his sword so that his technique is broken. Look for the next opening that presents itself.

"Ain bruch wider den bruch"

Item wann du zwerchst und dir och ainer mitt der zwerch wolt fornen für komen under dinem schwert an den halß so fall im nider mitt der langen schnyden starck uff sin schwert so ist es gebrochen nym die nächsten blöß die dir werden mag

As you see or sense his counterstrike, simply turn your sword upward and around so that you strike his blade with the long edge from above. Use the pommel to do this by pulling the sword, using the leverage provided by the grip.

"Against the low cut to the arms."

If you strike with an *oberhaw* and he parries this and goes high with the crossguard, and you do likewise and you both step in, then try with a lower cut. But if he wishes to cut at your hands from below, follow his sword down and push it down with strength using the true edge, then the cut is broken. At once look for an opening.

"Wider den schnitt unden zu den armen"

Item wann du ainem von oben eyn hawst und er das versetzt und fört hauch uff mitt dem gehultz unnd du och und lauffend mitt ain ander eyn so nim den undern schnitt und ob er dir den schnitt nemen wöllt unnder din hend in die arm so volg sinem schwert nach under sich mitt der langen schnyden und truck nider so hastu es gebrochen und such die bloß

If you strike with an *oberhaw* and he deflects this and goes high with the crossguard, and you do the same, and he tries to thrust the pommel under your hands to hit your face or to thrust the pommel at the breast, go well below with force, and the thrust (with the pommel) is broken.

Item wenn du aber hoch mitt der armen kümpst / und gener och also und lauff aber eyn unnd wölt er dich dann mitt dem knopff durch den arm under dinen henden under die ogen oder an die brust stossen so far undersich mitt dem knopff starck mitt den armen so hastu es gebrochen

After you have struck, your opponent tries to go high and step in, and you follow him up. He changes direction and tries to pommel you in the face. Simply go well below with force and the attack is broken. You can then counter with a low cut to his arms as you push him back.

If you are in the bind and he changes through with the pommel and alters his grip to halfsword, break it using the upper cut. And in the cut you can alter your own grip to halfsword and thrust.

In the bind your opponent can change through in various ways. He tries to come around you while using his sword as protection. In the changing, he switches his grip to half sword and attempts to thrust at your face or cut your throat.

Item wann du mitt ainem angebunden häst und wechselt er durch mitt dem knopff und felt in daß halb schwert daß brich schlecht mitt dem obern schnitt und im schnitt magstu in das halb schwert fallen und im ansetzen

Simply step to his outside and drop your sword so that it is well below his, and cut upward during the step. After the cut, you are well positioned to thrust at his exposed side. You can also place your sword on top of his arms and cut high in time with a step, or while pushing him back.

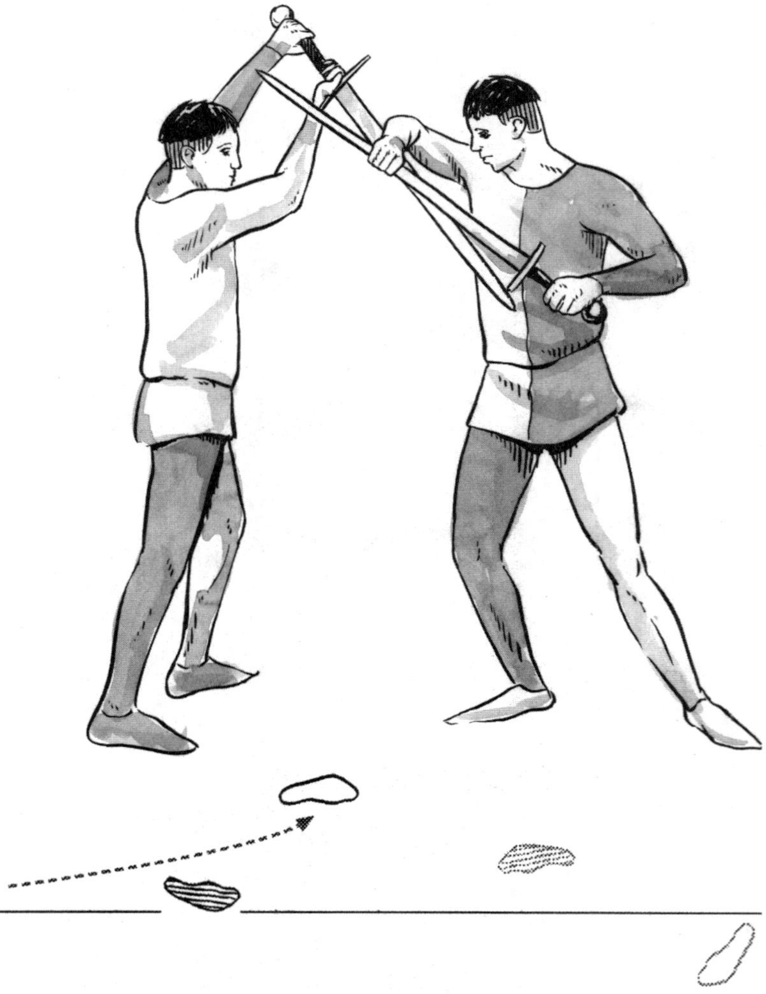

The Modern Practice of Swordsmanship

THERE IS BY now a veritable industry of people and products dealing with the practice of historical European martial arts. The central problem here is our approach to this activity. Do we try to re-create the fighting arts or do we try to understand them? Which is our primary goal? Both can most certainly be pursued at the same time, but our investigations require an emphasis, and this will determine the focus of our work. Groups ranging from medieval reenactment societies to organizations focused on researching every aspect of the historical fighting arts are helping increase our understanding of these skills.

IN PRACTICING HISTORICAL European martial arts, a term, which may encompass all weapons as well as unarmed styles used, we must first decide why we practice and what our goal is. This is paramount because it determines both how we should practice and what we will attain in the way of skills. Do we practice primarily as a pastime and to keep fit? Are we interested in participating in competitions, tournaments, or exhibitions? Is our primary interest research into the historical aspects, including equipment and techniques, or are we mostly interested in becoming skilled fighters? Of course, several of these aspects can be combined, but usually one predominates in our motivation for practice. It may change over time, but we need to understand why we do what we do.

THE REASON FOR this is very simple: *you become skilled at what you practice!*

THINK ABOUT THIS. It is evident that you always get benefits in such things as strength and agility. But there is also a question of how strong or agile you are. If you want to practice for the sheer joy of it, then you do not have any specific needs or goals to meet. But if you are interested mostly in understanding and communicating new knowledge to an interested community, you will have other needs to meet. Perhaps you must learn old German or Italian, depending on your focus.

AND IF YOU want to become a skilled fighter in the true sense of the word, you will again need something totally different. So, depending on the goals, we have different needs, and it is not until we have determined our goals that we can begin to practice a truly meaningful form of training.

BELOW ARE SOME of thoughts on how one can practice and what should be considered, gathered from the thoughts and writings of several prominent writers past and present.

1. Speed and strength are very important–between two equally skilled opponents, the stronger and more agile fighter will win.

2. In a real fight, with or without weapons, always resort to simple movements and techniques–that is how the mind works. Do not pay too much attention to very complicated or fancy techniques.

3. You cannot practice with all weapons. Find one or two you favor and work primarily with these. Otherwise, you will be incompetent with many instead of skilled with one or two weapons.

4. At each practice session be sure to practice unarmed, and perhaps include dagger fighting. This is always useful because there

are close-in techniques with other weapons.

5. Read everything you can find, but think before you accept anything as true. There is a lot of nonsense floating around today in the martial arts.

6. Fighting for fun and fighting for real are not the same, nor will they ever be.

7. Repetition is the name of the game. You need at least 50,000 repetitions of one movement before you can consider it to have become what is termed a "learned reflex." This is the level at which you really know how to do something. If you do some math on the time it takes to do 50,000 repetitions, and multiply that by the number of techniques you practice at one session, you will notice something quite depressing: You will die long before you have mastered them all. Again, be selective in anything but the basics.

8. It is better to be able to do a few things well, and understand the principles behind them, than to try and master all things at once and fail.

9. Practice daily. There is no substitute.

10. Do a lot of sparring using wasters and replicas. Without doing this you will not be able to perform your technique with any degree of efficiency. Use protection, but also strive for control and finesse with your weapon of choice.

11. Accept that you will most likely never use your skills in a real-life situation, so do not take things too seriously. If you take yourself too seriously, you might miss the important things that other people can contribute.

12. Have fun, and never take this too seriously. If you are not having fun, do something else with your spare time.

Physical Training and Swordsmanship

What do you need to do to become skilled in the use of a sword? First of all, you must accept that your body must be up to it. The entire thing will become more enjoyable if you can cope with the physical aspect of wielding a sword. We give some suggestions here on what to do and how to do it. The exercises are based on our own experience and are aimed at the swordsman specifically (not that they are original in any sense, but they work well for this purpose). Feel free to make any alterations you like; this is not gospel.

What we should be aiming for is a kind of "overkill" in regard to our physical ability. If we need 100 percent but can perform at 150 percent, that will make us perform better and reduce the chance of injuries. Of course, if you have a physical problem or are a senior citizen, practice within your limitations. But do not take those limitations as an excuse to not practice or as a convenient way out of having to make an effort.

AGILITY AND SUPPLENESS

There is no avoiding this, my friends: if you are stiff as a board, you will fence like one as well. I personally prefer to do yoga instead of stretching, since it will, at the same time (to some extent), mend your body and increase the endurance of your muscles. If you decide to take up yoga, try a few different styles first, and be sure to tell your teacher why you are taking up yoga. That way he or she can help you find the best method for you. Ashtanga (power) yoga is a good style, or maybe Iyengar-style yoga (but without props, if possible).

If you prefer to stretch, find a good book on the subject, like *Stretching Scientifically* (see bibliography for details). Avoid stretching designed for martial arts; go for gymnastic-style stretching instead. If you do not have previous experience, try to find a good teacher to show you the basics.

Remember that your body will become stiffer with age. Accept this and take it slowly if you are new to this. If you are older you can still become very agile; it just takes a bit longer.

STRENGTH

Here we are talking about something that is more generally available to everyone. We need strength to fence. Take a good steel replica and do full-speed strikes, thrusts, and techniques for 15 minutes without rest; if that is too much for you, then you need to do some strength-building exercises. For fencing we rely a lot on the upper body, specifically the shoulders, the forearms, and the grip. It is also in these areas that we are in open to injuries due to sprains or overexertion. However, using a sword is an all-body thing and it is best if you strive to balance the body in its development. Building strength in the legs is especially important for explosiveness in attacks.

ENDURANCE

You must have endurance. The level, however, depends on what you are aiming for. If you practice for the joy of it, take it easy. But if you want to fence with serious intent, you must train for serious endurance. Basically, you can choose between swimming and running. I prefer swimming since you have a lower risk of injuries to the legs, and you also have better benefits from overall muscle development. Try swimming for 30 to 40 minutes. As you get better, increase the tempo but not the length of time. And if you can, swim in the ocean instead of in a pool. You can also try running while wearing your armor.

SUIT UP AND run for no less than 20 minutes, the pace matters less. If you feel up to it, carry your chosen weapon as well. When you can do this comfortably you are getting somewhere! Remember that you need the endurance and stamina enough to fence for perhaps 30 fights in a row, each lasting maybe two minutes. That is one hour, and if your body gets tired, you will lose no matter how skilled you are technically.

The Interpretation of Historical Sources

When we strive to re-create (that is make anew, for there is no living tradition) the old European martial arts, we face several problems and pitfalls that put us at risk of falling into the trap of historicism. Since a living tradition does not exist, we are left with several diverse categories of source material to reconstruct the ideas, methods, equipment, and execution of fighting arts. But in both selecting and interpreting these sources, we face a large number of difficulties. Unfortunately, I must say that a large part of the work done so far on the European martial arts fails to address fundamental scientific questions. Of course, we may say that this search into the past need not be scientific, but I believe otherwise. We must live up to our own stated standards, at the very least. What I would like to do here is show some of the dangers associated with the material available to us, and some possible avenues around these pitfalls. Finally, I want to point out the absolute limitations we are faced with and around which there is no scientific way at this time.

From the outset, it is of paramount importance to understand and distinguish between a hypothesis (an idea you have and that you try to ascertain the truth of) and what is within reason to be considered as proof or to have been proved. The latter I dare say is *very* rare in historical sciences, but used all too frequently by European martial arts proponents.

THE WRITTEN MATERIAL

There are written accounts of battles, duels, and the like, as well as all the manuals describing techniques and principles. These include illustrations of arms, armor, and combat. If we begin with the account of combat, regardless of whether it is a battle or a duel, we first have the eyewitness problem. That is, we cannot actually know whether a person was an eyewitness to the event or whether he was repeating hearsay and including himself among the spectators or participants. To establish presence we need an independent corroborating source.

Another avenue is to take the subject we find interesting (say, arrows piercing armor) and go look for other medieval sources that mention the same thing. The crux of the matter is that we must try to establish their dependence (has one influenced the other?) and, if so, in what sense. Also, if it is a modern translation of medieval German, for example, does the translator understand the subject in question? If not, there is a danger that several important terms are actually misinterpreted. Also, we have to question the writer who actually wrote the book. What was his aim, for whom did he write it, and what was the historical context of his time? Did he perhaps hold a grudge against another teacher? Did the book turn out as he wished? Perhaps the illustrations were wrong, but due to lack of time or funding they were kept, or maybe a poor illustrator was the only choice for various reasons. Another common problem was the use of similar engravings or carvings to save time in printing, i.e., printing blocks were used from other, already printed books.

All these questions must be answered to understand the text properly. It is not possible to say that a fencing manual automatically is either representative of the master of the school in question or an adequate instruction on how to use the weapon. We must examine the material from all angles, and then take our results and try them out with swords ourselves. So the written material is open to question from many angles. How about the illustrations, then?

TO BEGIN WITH, the use of, or rather the lack of, perspective prior to about 1450 gives serious problems in understanding body position and angles in illustrations. Second, it was and continued to be a convention to show the blade with the flat toward the reader in order to show the sword–not because this is necessarily how the sword was to be held. Also, there are the standardized legs ("banana legs") that look the same with very little modification up to about 1500, regardless of the situation depicted in the picture. These simple facts give us serious problems with interpreting the illustrations of various techniques and even equipment used, since we cannot trust the picture directly, but must interpret it in the context in which it was created. Maybe the artist was incompetent. We do not know. Therefore drawings cannot be taken as gospel.

WITH PICTORIAL MATERIAL there are a few simple rules in interpretation. These are not new, nor do I claim any originality in any way.

LOOK AT HOW the hands are turned before trying a posture. Simply place the hands as in the picture. Pay special attention to the thumb and the back of the hand. The reason for this is that even though medieval masters could not draw perspective, they showed you how to hold the blade by indicating the position of the hands. This will tell you how the blade is to be angled or held.

Note which leg is leading. Do not bother to assume the exact position of the legs as in the picture, just look at which leg is leading. Also note if the legs seem to be crossed or very close: this may indicate a special situation. Then assume a stance that feels comfortable and that offers speed and stability and resembles the illustration as closely as possible.

Note the position of the arms. Are they bent or straight? Are they close or far away from the body? Make sure that you take account of the position of the hands here as well.

PUT IT ALL together in one stance and play around a bit. Most likely if you couple this with the most common strikes and thrusts, you will get a pretty good idea of how an image was conceived in the author's mind. It's simple but works well. As advice, never even try to perfectly duplicate a position or movement from an illustration because it never works very well if put to the test. (After 1500, however, there was an upswing in the skill of the artists involved and the illustrations were more realistic.)

THE ARTIFACTS

FORTUNATELY WE ARE blessed with yet another category of material to work with: preserved archaeological and historical artifacts. These can be weapons, armor, or osteological remains. All these tell a direct tale of the use of medieval weapons and the injuries caused by them. Unfortunately, not even here can we accept the material at face value. First, we must establish whether the weapons and armor had actually been used in battle. If not, then any damage to the items or lack of it cannot be used to discuss how the swords were used. The fact that the sword is in a museum is not enough.

THE OSTEOLOGICAL REMAINS pose an even greater problem. Here we must be very careful with the conclusions we draw. The material is interpreted by archaeologists before the public gets its hands on it, but archaeologists are very seldom knowledgeable in the field of arms and fighting and often draw faulty conclusions. Unknown factors play a part as well. For example, much has been made of the material from the battle of Wisby. What most people do not know is that the material was collected and later categorized. Very few individuals were assembled from the material, and bone specimens needed for age and other analyses were removed and placed together. Also, all material showing battle trauma were set aside, but the distinctions between individuals fail, which presents problems in the interpretation of the material. This means that today, when we try to understand how the damage was distributed (the famous percentage lists that everybody quotes), we have no way of knowing how much damage was actually done to one individual! The material was dispersed when collected and there is no way short of a complete DNA test to put an individual back together. And if we cannot study an individual, the material becomes next to useless. (The famous argument, "go for the leg and then for the body," cannot be sustained with this material, at least not in any scientific way.)

So, we must know and understand how the material has been collected and we must be able to understand the methods used before we can have any chance of establishing the relative value of the material for our investigation. This is very often not the case today and has resulted in the floating of some rather outlandish theories.

THE WAY FORWARD

Speaking from experience, I would like to make a few points for future studies of the medieval martial arts.

Know your subject. If you use historical or archaeological material, then you must understand the basis for these sciences. This will enable you to use the material wisely instead of just repeating what others have probably already said, including their mistakes.

Accept that we have definite limitations to what we can know of the actual and cognitive worlds of the medieval warrior. There is nothing more ridiculous than saying, "This is what they thought." Rather say, "Perhaps this is a useful model for gaining a better understanding of how they thought."

Be critical of both the material and those who use it. There is a deplorable tendency today to regard the historical manuals as the word of the Lord Almighty, and to think that only one book can claim that status. This is a sure way to misunderstand the medieval martial arts.

Always try to correlate the historical material with archaeological material or artifacts in museums. Words can lie, but a sword is a sword. To answer our questions it is important to examine materials in light of other material categories.

What is an Edge?

by Peter Johnsson

THE IMPORTANCE OF SHAPES AND DIMENSIONS

THE SWORD HAS changed shape through the ages according to the demands and possibilities of the time and surroundings. Such factors as the way wars are waged, the varying techniques of swordsmanship, the nature of available armor, the quality of materials used, the development of the metalworking crafts, and even the organization of society and the economy have influenced the shape and type of swords used in any given period of history. What we see cannot simply be described as a linear evolution from the crude to the advanced; the design of the sword is always in harmony with the needs of its age.

THE SWORD HAS its specific shape in order to enable the wielder to use the edge and point as effectively as possible. Despite this obvious fact, it is very easy to misunderstand and misinterpret the meaning and importance of the properly balanced weapon and its sharp edge. What does it mean when we say that a sword has a good balance? Just how sharp is the edge on a sharp sword? What is needed for a sword to be an effective weapon in a violent confrontation?

THE FUNCTION OF the sword is decided by the shape of the blade, the cross-section in different parts of the blade, the gradual change in flexibility along the blade, its ability to absorb and distribute vibrations and shocks, and the form and character of the sharpness in the edge and point.

THE HANDLING CHARACTERISTICS of a sword are determined by its total weight, the place of its point of balance, and the distribution of the mass in hilt and blade. The total weight of a sword can vary without moving the point of balance. The distribution of mass may be changed without necessarily affecting the point of balance or it can be manipulated to achieve just that. Each of these three elements–weight, mass distribution, and point of balance–are put in tune so that the sword may follow the will of the swordsman with a minimum of effort. Many of the factors above interact in subtle ways, sometimes reinforcing each other but at other times acting in opposition.

THE RESULT OF functional and handling characteristics determine how well a sword will cut and thrust. This is a matter not only of the depth of cuts and thrusts, but also how easily the sword will deliver an attack.

A SWORD DESIGNED primarily for thrusting will have such heft that a minimum of force is required in order to control the movement of the point. This is done by having small mass in the point compared with the area at the base of the blade close to the hilt. The placing of the center of gravity is important to the accuracy of the thrust. It takes more or less force to make the point deviate from its path, depending on the distance between the point of balance and the hilt.

WEIGHT AND MASS distribution influence the amount of energy needed to accelerate the sword during cutting. The total weight and the amount of mass in a specific distance from the center of movement (the hand) affects both the speed and the force of the strike. A lighter sword can be balanced farther out toward the point without making it excessively blade-heavy. A sword that is too light will not generate enough living force to cut through a difficult target. The heft of a sword should enable it to bite deeply and thrust with authority, yet remain easy to control. Too much mass placed in the extreme ends of a sword, its point and pommel, makes it sluggish, less exact, and dull in delivering thrusts and cuts. A small amount of mass in the point area will make the sword quicker in handling. The point and cutting area of a blade must not be so slim, however, that it cannot withstand the shocks and vibrations caused by impact.

A sword primarily intended for cutting usually has a more even distribution of mass, while a thrusting sword has the mass concentrated closer to the hilt. The center of gravity (the point of balance) must in both cases be adjusted so that the total weight and distribution of mass in the sword work together for a maximum efficiency. The balance of the sword and shape of the blade also determine its tendency to vibrate. As a blow connects a sword will ring and vibrate. Severe vibrations will make the sword deliver less energy to the target.

There are areas where vibrations are dampened, called nodes of no vibration; energy delivered here causes less vibration in the sword. If a sword is struck or delivers a strike close to a node, the sword will be perceived as stiffer. These nodes are of great importance for both cutting and thrusting swords because they influence the cutting ability and, to some extent, the character of a slim blade used in fencing. The manipulation of these nodes is an important tool in the making of a sword, as important as the overall sizing and balancing of the blade. Placing the nodes is decided primarily by the shape of the blade and is adjusted by the dimensions of the hilt.

The size, weight, and, therefore, the handling characteristics vary among swords of similar type. When handling original swords in collections, one can easily get the impression that differences in weight and proportion may depend on whether the weapon was intended for the battlefield or for civilian use. This may be something to consider when we study illustrations in fencing manuals, or draw conclusions from well-preserved swords in armories and museums.

HISTORICAL EXAMPLES

The longsword depicted in fencing manuals where halfsword techniques are illustrated is of a specific type. It is wide at the base of the blade with edges that taper straight to an awl-like point. The part of the blade used for cutting is not very wide. A pronounced ridge gives a strong diamond cross-section. Occasionally a short fuller is depicted at the base of the blade. In other period manuals, swords are shown that have little or no tapering in the width of the blade. The points of these swords are shaped like spades. In manuals where these swords are shown, we usually see no halfsword techniques. We can find examples of both of these swords in museums and armories. Both types definitely have sharp edges, but the character of the edge is different. This should come as no surprise, since the first type is mainly used for thrusting, while the other is primarily for cutting.

To learn about the nature of the sword edge we can study well-preserved historical originals. These are functional objects made to meet a specific demand in a specific situation. A comparative study of medieval illustrations and historical swords surviving in collections is rewarding. The swords we recognize as being used in halfsword techniques have a design that emphasizes thrusting. The edge is sharp and can surely deliver devastating cuts, but perhaps not as effectively as the broader blades designed primarily for cutting. Even if the angle of the sharpest part of the edge is more or less the same on both types, the two blade types allow a different purchase in grasping them. The sword designed primarily for thrusting is narrow with a raised midrib, while the cutting sword is wide and flat. The wide, flat sword has a greater cutting capability than that with a narrow blade and is less inviting to a grasping hand. The cutting blade is also more flexible and may, therefore, be less suited for the spear-like movements of the halfsword. We see that impressions gained from the study of the original sword reflect on our interpretation of fencing techniques.

THE SHAPE OF the edge and its sharpness complies with the intended function of the sword. Blades with different cross-sections carry edges of varying character. Rhomboid cross-sections and lenticular, hexagonal, or hollow-ground blades all allow different types of sharpening of the edge. The outermost fraction of the edge is often of a slightly larger angle than that of the main edge bevel. The transition between edge face and cutting sharpness follows a gentle curve so that a cross-section of this outermost part resembles a very pointed Gothic arch. An edge like this allows an aggressive bite while being sturdy enough not to break or chip too easily.

Figure 1.

Figure 2.

THIS GRIND IS famously present on the impressive Japanese swords, but it is also found on all swords from the Viking age. These often have a markedly convex edge bevel that gradually transforms into the sharp edge, allowing an effective mass and sharpness for tremendous shearing blows. In Figure 1 you see a detail of a sword found at Fullerö, Sweden, which is stored in the basement at the Historical Museum in Stockholm. Note how well the pattern-welded section aligns with the outline of the fuller. The sword is very well preserved, and you can still see faint streaking in the surface of the carbon-steel edge. The blade retains a great deal of the original sharpness despite its age.

SWORDS DATING FROM later periods with blades designed for cut and thrust have variations of this type of edge. We find this grind on the broad-bladed weapons, commonly occurring from the 12th up to the mid-14th century. Note the shape of the blade in Figure 2. This is the point of a war sword from the 12th century, stored in the Army Museum in Ingolstadt, Germany. The cross-section has a rounded, rhomboid shape at the end of the fuller, and it gradually changes into a more lenticular shape toward the point.

The longsword as such exists in a variety of shapes that often emphasize either the cutting or thrusting capability of the weapon. Weight and size may vary a great deal between individual weapons. A longsword of small but beautiful proportions is on display at the Deutsches Klingenmuseum in Solingen, Germany. A very elegant and slender weapon of exceptional quality, it gives the impression of having been made for a child of noble birth. The cross-section is of a very full lenticular shape. The point is very pronounced and stiff with a marked midrib. This is a very fast weapon probably intended for thrusting and slashing cuts in soft targets. A sword like this will not deliver powerful cleaving attacks due to its lightness and the rather sturdy section of the blade. It is a perfect example of a civilian weapon for self-defense. The total length is just over 1 meter (39 inches) with a blade length of 816 millimeters (32 inches). The weight of the weapon is only 822 grams (29 ounces).

Longswords with stiff blades designed for both cutting and thrusting often have a pronounced midrib with a diamond-shaped section. If the cutting capability is more important, the sword will have a slightly flatter and wider blade. Swords like these were in vogue during the latter part of the 14th and 15th centuries, but occur in both later and earlier periods. An example of the narrow variety is the fantastic longsword at the Bayerisches Nationalmuseum in Munich. The blade has a strong diamond-shaped cross-section with perfectly flat edge bevels that continue straight into the edge. It is in absolutely perfect condition. It has been used, though. There are some very slight irregularities in the edge that suggest resharpening after use. An example of the wider type of longsword that may yield better cutting performance is a well-preserved weapon in the Marzoli collection in Brescia, Italy. The cutting exercises in this book were done with a replica of this sword. It combines excellent handling characteristics with good cutting performance. A sword of this type may perhaps not cut as deeply as a wide-bladed sword of war, but it will be extremely responsive in dedicated fencing. This sword may be one of the best surviving examples of the type of sword so often depicted in medieval fencing manuals.

Many rapiers have a lenticular, rhomboid, or hexagonal cross-section where the outermost part of the edge toward the point is sharpened almost like a filet knife. A thin and extremely sharp edge can work well with this design without being submitted to too great a strain in use. Such a blade will not cleave more resilient materials like heavy leather or bone due to the small mass of its point, but it will still have a very good cutting ability in slashing cuts against unprotected areas.

Figure 3.

TO RIGHTLY UNDERSTAND the quality of an individual sword we must consider its intended use and function.

THE EDGE AND the point are the two most vital parts of any sword, yet they also quite fragile if abused. The design of the blade and the grinding of the edge reflect the intended use of the sword. There is a great difference between the realities of the battlefield and the civilian duel throughout the ages.

OCCASIONALLY DAMAGED BLADES survive with marks from use still intact. In Figure 3 we can clearly see damage in the edge of a single-hand sword from about 1500. The original is in the Deutsches Klingenmuseum, Solingen. The blade has a thin, diamond-shaped cross-section with a very slight convexity in the outer few millimeters of the edge, creating an edge angle of less than 40 degrees. At a distance of 2 millimeters from the edge the thickness is about 1 millimeter–slim dimensions, but not unusually so. The cutting part of the edge is in fact not unlike that of many kitchen knives, with the only difference that the convexity of the edge bevel leaves the edge slightly sturdier. The nicks in the edge were caused by another blade meeting it at a slanting gliding angle during a parry of some kind. The damage to the sword shows that the two swords never met in rigid blocking parries, not even at the forte of the blade.

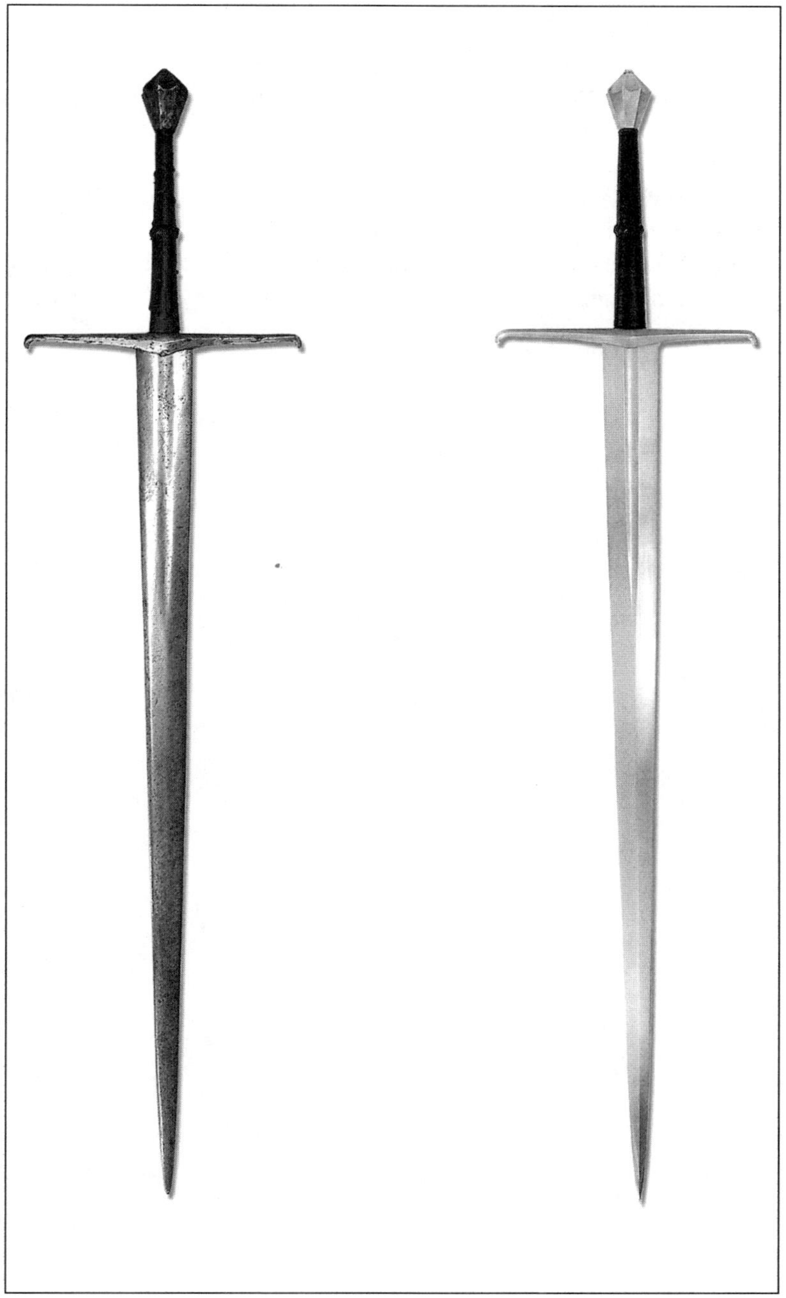

This is the sword we used for the techniques and the test cuttings, a spadona *(a word simply meaning a large sword in Italian) dated from 1450, currently in the Collezione Marzoli in Brescia, Italy. The blade is from Passau, Bavaria, while the hilt is of Italian type.*
The sword is very fast and yet cuts extremely well and may be said to represent the type of sword indicated by Fiore Dei Liberi, Ringeck, and others. This is a longsword well adapted for the techniques in Ringeck's manual.
The replica was made by Peter Johnsson for Albion Armourers.
Total length: 1,160 millimeters (46 inches); blade length: 995 millimeters (39 inches); weight: 1,470 grams (51.5 ounces); point of balance: 100 millimeters (3.9 inches) from the crossguard.

The damage was most likely caused by the two swords meeting at an angle of around 30 degrees in deflecting, pushing movements. No nicks in this sword are particularly deep, most being less than 1 millimeter deep.

Other blades show that more severe damage could be sustained in the edge, perhaps a silent statement of the desperation in the fight. We may perhaps assume that the schooled violence of the civilian duel and the havoc of the battlefield leave different marks for posterity.

1. A small longsword (for a youth?), 0.822 kilograms (1.81 pounds). Late 15th century. Deutsches Klingenmuseum 61.154 (Solingen)

2. Longsword, late 15th century, 1.47 kilograms (3.2 pounds). Bayerisches Nationalmuseum W.872 (Munich)

3. Longsword, mid-15th century, 1.4 kilograms (3.2 pounds). Collezione Marzoli G4-664 (Brescia)

4. War sword, 12th century, 1.87 kilograms (4.1 pounds). Bayerisches Armémuseum A3537 (Munich)

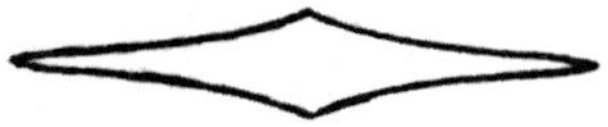

5. War sword, 1490–1500. Belonged to Svante Nilsson Sture, 1.8 kilograms (4 pounds). Västerås Domkyrka.

6. Falchion, c. 1500, 1.1 kilograms (2.2 pounds). Deutsches Klingenmuseum 69.w.4 (Solingen)

7. Spada di San Maurizio, 1200–1250, 1.33 kilograms (2.9 pounds). G.25 Armaria Reale (Turin)

8. Sword type XIIIb, 1300, 1.4 kilograms (3.2 pounds). Stockholms Medeltidsmuseum (Stockholm)

9. Sword type Xa, early 13th century, 1.17 kilograms (2.6 pounds). Historiska Museet, Stockholm (Stockholm)

10. Large warsword, 1480–1500, 2.025 kilograms (4.5 pounds). Skokloster Slott (Skokloster)

Note: The profiles are drawn to 1:1 scale and show the cross-section of the blade node.

How to Sharpen a Sword

by Peter Johnsson and David Lindholm

SHARPENING AND MAINTENANCE OF YOUR SWORD

When a sword is used for test cutting, it will be subjected to varying degrees of wear that sooner or later will dull the edge. It is good practice to frequently check the edge to see if there is any need for maintenance. An edge that has lost a fraction of its bite may easily be restored with a hard whetstone or a piece of very fine emery cloth secured on a flat piece of wood. If the sword is kept in good repair and polish, there will seldom be any need for more involved grinding work. A slight touch with a hone now and then will keep the edge cutting during many sessions.

If the edge has been chipped or has lost its proper cross-section from being sharpened many times, it will sometimes be necessary to do more involved work. The cutting edge and its geometry can be restored either by filing or grinding, depending on the hardness of the metal. If you decide to change or reestablish the shape of the cross-section, first make sure that you have a good mental picture of what you want to achieve. Always have a clear idea of the purpose of your work. Think about what type of targets you intend to cut and adjust the edge of your sword accordingly. Swords of different sizes and types need slightly different edge shapes. Keep in mind the nature of the edge and be careful as you remove excess material or establish a sturdier edge. Become familiar with how the gradual change of angle and thickness toward the cutting edge feels between your fingers. It is generally only the last 5 millimeters (0.4 inch) or so behind the edge that need reworking to establish good edge geometry.

This kind of light stock removal can be made by the help of a belt

Here are the tools. From the left; oilcan, emery cloth in several grades, metal files of several shapes and grades, a piece of wood to wrap the emery cloth around, underneath that a piece of Scotch-Brite, a whetstone in a small can of water or oil, and a leather strop.

Using the whetstone, place the sword on a piece of cardboard at the edge of a table or workbench. This gives you good control of the angle as you work. Make sure that the stone is wet. You can also oil the sword first. Work in small, firm movements toward the edge.

Oil the blade well before using the Scotch-Brite. Scotch-Brite is great both for removing marks as well as doing the next step in the sharpening process. Continue to work toward the edge and the point.

When using Scotch-Brite, make sure that the angle, as when using a whetstone, is not too great. Easy does it.

Finally, the strop. Do no more than five strokes on each side of the edge. If you do more you risk dulling the edge you have labored over. Again, move toward the point of the sword.

grinder or even a file, as most European-styled swords have a hardness that is just soft enough so that a new fine-toothed, single-cut file will bite. Working with a file is slower but less risky. Be very careful when using power tools: a short moment of inattention may easily ruin the blade and/or lead to more or less severe personal injuries. The edge must never be overheated so that the temper of the blade is damaged. Frequent cooling is necessary. Even if the body of the blade is still only moderately hot, the cutting sharpness may already have reached a damaging temperature. It takes a light hand and good control to do this well. To avoid the risk of overheating altogether it may be a good idea to install a water hose with a nozzle that creates a cooling mist where the blade and contact wheel meet. Be aware that long working sessions at a water-cooled belt grinder might cause problems with rust on the blade and hilt of your sword. Wipe the blade dry frequently and rub it with a piece of oily cloth. You may also protect the hilt with plastic.

In historical illustrations we can see how sharpeners and grinders worked at wheels that were water cooled, either by having the stones partly immersed in water or by sprinkling water from an overhead pipe. If you can find a good sandstone wheel and learn to use it, this may be an ideal tool for reshaping the edge of your sword.

Grinding stones and belts will leave a surface that may be coarser than the overall surface of the blade. More or less touch-up work with fine emery paper may be required if you wish your blade to have a uniform polish on all surfaces. The surface of the sword can be polished with a pad of fine Scotch-Brite to establish a good finish before the very last part of sharpening. The use of light oil for lubrication helps in achieving a nice and smooth polish. There may be difficulties in reaching a good final sharpness if this is overdone because the Scotch-Brite pad may harm the carefully shaped edge. To avoid this, the edge can be leaned toward the face of the supporting working table so that the thin line of the edge is unharmed.

To put a biting sharpness on the sword, the edge is worked at slightly increasing angles with honing stones or emery cloth secured to a flat piece of wood. Emery cloth is preferably lubricated with oil, while stones can be used either with oil or water. Liquid soap can be added to the water to keep the blade from rusting.

Start with a coarser stone and work in sections along the edge. Watch out for a burr that forms when the edge starts to fold. You can feel this as a ragged edge if you carefully follow the surface of the blade toward the edge on the opposite side from which you are working. This is a good sign that you have completed the honing on that part of the edge. Move on the next section. Turn the blade and repeat the procedure from the other side of the blade so that a burr starts forming on the first side. Always stop working on a section when the burr has formed and continue working until there is a burr along the whole section that is to be sharpened.

Use a finer stone at a slightly less acute angle and form a burr on both sides as before.

The final sharpness is achieved by polishing the edge on a flat piece of leather, a strop. This should not be overdone because the fine edge might become rounded. A couple of strokes on each side are enough to produce a hair-splitting sharpness.

It is common not to sharpen the 15 to 20 centimeters (approximately 6 inches) closest to the crossguard to an acute sharpness on most longswords. The sharpest part of the sword, the last two-fifths of the blade toward the point, will do most of the cutting. The challenge in sharpening a sword is not so much producing an acute and precise edge, but doing this while keeping the cross-section slim but sturdy. Cutting performance must be balanced by edge retention.

If the blade has a cross-section that is too thick behind the edge, it will drag in the cut. An edge that is very thin might cut well but not stay sharp for very long. There is not one specific edge angle and cross-section that will function on all types of swords. The shape of the edge must be adjusted according to what the sword is intended to cut. The dimensions and size of the blade will also influence the nature of its sharpness. Lastly, the quality of the heat treating, hardening, and tempering of the blade will determine what kind of edge the blade can hold.

Generally speaking, the main edge bevel of a cutting sword may vary between a total angle of 10 and 30 degrees, while the sharpening of the edge is shaped to an angle that varies between a total of some 30 to 50 degrees. The transition between these angles is, as a rule, a gently curving line. Swords that are intended to cut lighter targets may have a very slim edge profile, while swords for heavier cutting need a sturdier section.

A thin edge will slice with almost no effort, but requires more maintenance. A sturdier edge will last longer but makes less elegant cuts.

Once a good functional edge is established on a sword, it takes little effort to keep it in good repair. Do not overdo this, however. Avoid sharpening the edge at too blunt an angle. If this is repeated often enough you will soon have to reshape the main edge bevel. Use a fine stone with a light hand to keep the edge sharp, and the sword will be a faithful tool during many pleasant training sessions.

Test-Cutting

First of all let me state that test-cutting is a must, period. Let me explain why. When you practice strikes or techniques with a steel sword, a waster, or whatever you use, you do not strike to actually hit a target most of the time. This will affect how you balance your body during the movement and how you react to the impact. There is also a vast difference between actually trying to inflict damage with a sword and merely slapping something. I have been at many seminars where people fail to cut through a cardboard tube (an easy target) even though they have practiced swinging a sharp sword for years without actually striking anything. They can move the sword well, but not with intention and real force. The sad truth is that no sword will cut for you–you must do for yourself. And just as with any martial art, there is a world of difference between waving a weapon around in the air and using it for real. So test-cutting is important, but how do you do it in the best way?

First, try to incorporate test-cutting into your regular practice and try to do it with a group. There are several reasons for this. It's more fun, and if your strikes are bad in some way, your friends can offer invaluable observations of how your body and sword move. While perhaps not everyone can practice cutting at each training session, try to do at least 100 strikes every month, preferably 200 or 300.

Now for the sword. You need a high-quality steel replica–do not waste your money on a cheap and inaccurate replica. Buy a sword that is meant to be both sharpened and used. There is no need to be totally anal on the issue of its similarity to an existing sword in a museum. There are so many swords preserved that you can justify almost any design. But by purchasing from a reputable craftsman, you will get a well-balanced sword that is properly tempered and handles well. And bigger is not better; you need a sword that is right for you, and that means that your strength and height should be taken into account. If you are sure to inform the seller what you need the sword for, he will give you sound advice. A good sword will cost, but it is possible for several people to pool their money and buy a sword for the study group. Then everyone who put his money in can use it. This is a good system that we use ourselves. (In the online resources section, there is a list of sword makers that I personally like and know make first-class products.)

Start easy on the cutting; do not try to cleave a cow on your first attempt. When I teach people I let them progress from water-filled plastic bottles or milk cartons to rolls of soaked newspaper. These can be modified by changing the thickness, but wet paper is a good target in that it allows you to make mistakes without damaging your sword. It also lubricates, an important factor very often overlooked when discussing how a blade cuts a live body. Lubrication improves the cut by about 30 percent. It is a good idea to make sure the target is lubricated when you test cut because any dry material will do considerably more damage to your edge. At a later stage you may roll the wet paper around a plastic tube. Get one of the same diameter as the bone in the arm or thigh and make sure that the walls of the tube are no less than 4 millimeters (0.2 inch) thick. You will have to experiment to find a type of tube that does not shatter on impact (bone does not do that). We have found a green tube called Aquatherm, made in Germany, between 4 and 15 millimeters (0.2 to 0.59 inch) thick with a hard, wax-like feel to it; this works very well. Naturally, you can substitute other material for the core, but avoid wood or old bone, since both are useless. With this layered target you can simulate everything except the head as a target for cutting. It is possible to use other targets as well, such as cardboard tubes, bamboo, plant stalks, or old pieces of armor.

You have your target, now it is time to rig it. One thing you must never do is place a target on a chopping block, table, or similar rigid object. When deciding how to mount your target, you must understand that a human body that is struck by a sword moves with the strike, which will dissipate some of the energy accumulated in the sword and delivered in the strike. A rigid support will not damage the sword, but it is very incorrect because the block will respond too much in the strike and give a much better result than in real life. (There is a reason you split logs propped up on other logs, you know, and it's not the reach.) We tried to strike a hole in a 3 millimeter (0.1 inch) plate lying on a chopping block with a small replica of a medieval hand axe with no problem. In real life, if a person had a 3 millimeter plate on the body–forget it, there'd be no effect from such a strike. So how do we design a good, cheap, and efficient target?

I prefer to use a solid post in the ground as a base and then take the target (which should have about 15 centimeters [approximately 6 inches] of pipe or spare paper sticking out) at the end. I then simply duct tape it to the post. That way it will move with the strike and not provide too rigid a target. Another favorite is to hang the target by a sturdy cord from a beam or branch, which demands very good technique to cut through. You can also attach a post to a 40 x 40 centimeter (16 x 16 inch) plywood platform and then do as above with the duct tape. That way the target will be even harder to cut well. The point is this: a body moves *a lot* when struck with a sword, which dissipates a lot of the impact energy. This must be simulated when test-cutting.

But in order to get the most out of test-cutting, you must practice. I like to place several targets around me in a pattern so that in a single sequence I can execute several attacks while having to change the angle to the targets. This allows me to work on my balance in a strike.

This is my preferred kind of target. You see the pipe in the center and the tightly rolled paper, well soaked. The duct tape holds it together. Remember to remove any staples in the newspapers because they might damage the edge. This strike was made with the Brescia sword Peter Johnsson made, and as you can see it works well. The diameter of the tube here is 32 millimeters (1.25 inch) and 5 millimeters (.20 inch) thick, wrapped with 40–50 millimeter (1.5–2.0 inch) paper and two final layers of duct tape.

The Use of Armor

Even though this portion of Ringeck's manual does not contain any specific instruction on how to fight in the heavy armor of the 15th century, we would like to include a few comments on it. On the right you see a German style full plate armor from the second half of the 15th century. It covers the entire body and provides more than ample protection. Wearing this kind of armor renders any striking attack with a longsword useless; at worst you will hear church bells. The edge will not penetrate the steel, but a thrust is another matter. Note that the protection is concentrated to the outside of the extremities, leaving the inside unprotected. This is to facilitate movement and reduce the weight.

As you look at the picture, you see that the angled attacks that would be made from the *winden* and *hengen* point most of the time to the openings at the inside of the arms or legs or at the unguarded area at the armpit. The halfsword thrust does not, as has been suggested elsewhere, penetrate the plate. What you can do is move in close with your sword length reduced to prevent your opponent from attacking and thrust with greater accuracy. This means that you can move, but he will have problems doing so due to the confined space.

Most of Ringeck's techniques can be used in armor, but not all will be effective even to a small degree. The difference in fighting in armor is that you can make more mistakes, but every movement is twice as hard to perform, so you will get tired a lot faster. The techniques tend to become less circular and more linear, and the use of counterstrikes at the opponent's blade decrease; instead, try to counter directly to the body. Wrestling and close-in fighting become considerably more dangerous because you cannot move as easily, and a hold or push becomes harder to counter. Naturally, a full suit would not have been worn by foot soldiers of the time, but primarily by mounted troops. As for weapons, the pole arm in all its forms was the primary arm for most foot soldiers. It is safe to say that the longsword fighting of Ringeck and his predecessors arose from the use of the longsword by men in armor, but as it developed it became more of an unarmored style, with other types of weapons supplanting it on the battlefield.

However, the longsword developed into something new in the judicial and civilian arena, so perhaps we should not draw too distinct a line between the earlier battle style and later civilian style of fighting with the longsword.

Full Gothic plate from the second half of the 15th century. The armor must have a tight fit. Very often today reenactors and others wear Gothic armor that is much too large for them. The halberd was the dominant weapon on battlefields, along with the crossbow. These weapons were fearsome in combination with a full suit of armor.
Armor made by Albert Collins–Via Armorari in Sweden

Glossary

Abnemen–To free yourself from the bind or to make an attack from the bind.

Abrayßen–To follow your opponent's weapon or his hands, usually a downward motion with the hilt.

Abschneiden/Abschnyden–To cut over the arms from below or from above, usually done with the long edge.

Absetzen–To set aside. The expression denotes the idea of setting an attack aside, usually a straight attack such as a thrust against the lower openings. It is also possible to deflect and catch a strike with the absetzen, using the flat of the blade to catch the incoming sword, which then slides down to the crossguard or skids off completely.

Abwenden–To "ward off" a blow as with a deflecting strike. Literally "turning aside."

Alber–Fool's guard. One of the four primary guards. A low middle stance with the sword point down to the ground with your arms outstretched, leading with either leg but usually the right. The hilt is held low and the blade at approximately 45 degrees. A very useful guard, if somewhat limited in its offensive capabilities, it provides excellent defense. It looks very exposed and thus foolish, while in fact it is very easy to launch effective counters from it especially if combined with a *durchwechseln.*

Alter schnitt–The old cut. A cut over the arm of the opponent when he has repulsed a *nachrayßen* (traveling after). To move in to cut on the opponent's extended arms as an action following the completion of his strike.

Am schwert–At the sword. Attacks made while maintaining contact with the your opponent's blade.

Anbinden–The weapons in contact. The making of the bind.

Ansetzen–To attack a specific body part, often with the point.

Ausser Nym–The outer/outside taking. A nachrayßen, that is a thrust under the opponent's weapon to the lower openings from the outside.

Bedebern/Bedebren–To defend with stabs or strikes.

Binden/Band–The contact between two swords.

Bloßen–Openings. The four openings are areas to aim at in combat, the first opening is the opponent's right side, the second opening is his left side above the belt. And the other openings are the right and left sides below the belt. Sometimes called windows.

Bruch–Break. In swordplay, the counterattack that breaks the opponent's attack.

Buffel–Ox. Insulting term for a fencer with little skill who relies on his strength and aggression alone to fence.

Drey hew–Three strikes. An *unterhaw* from the right, followed by an *unterhaw* from the left, and a powerful *schaytelhaw* to the head.

Drey wunder–The three wonders. The three components of the system: the strike, the cut, and the thrust.

Duplieren–To instantly follow up a parried long-edge strike with a strike through a swift crossing over of the arms. The left hand that guides the sword pommel goes under the right forearm in this position and your sword is between the weapon and body of the opponent, hit with the long edge in the head. It is a kind of winding action.

Durchlauffen–To run through. This is a term meaning that you run in under your opponent's attack when he is very high with his sword. Your own blade is positioned differently depending on situation, but it is often hanging down across your back while you hold it with your left hand. You can wrestle or step in to cut with the *durchlauffen.*

Durchsetzen–To push through. A timed thrust from above or below that passes between the opponent's outstretched arm and his body.

Durchwechseln–Changing through. The motion of evading contact with the opponent's blade as you strike or thrust (e.g. changing the line of attack). To change into another opening with the point, with or without a bind. From the bind it becomes very important if the opponent binds hard or soft.

Einlauffen–Running in. To duck in under the opposing weapon or employ closing and entering techniques. This can also mean that you rush in with speed and force to close the distance with the opponent.

Fechtbuch–Fencing book. A German manual on fighting techniques and methods, particularly swordsmanship, (plural *fechtbücher*) among the more famous are those by the masters: Johannes Liechtenauer's of 1389 (by Hanko Döbringer), Sigmund Ringeck c. 1440, Hans Talhoffer of 1443, Peter von Danzig of 1452, Paulus Kal c.1460, Johannes Lecküchner ("Lebkomer") of 1482, Peter Falkner of 1490, H. von Speyer of 1491, Jörg Wilhalm of 1523, Andre Pauerfeindts of 1516, and Gregor Erhart from the early 1500s. Medieval Italian fighting manuals include those of Tarcirotti c. 1400, Fiore dei Liberi from 1410, Boris Ferres of 1428, Fillipo Vadi c. 1480, and Pietro Monte of 1509. There is also the Spaniard Diego de Valera's book c. 1490.

Fehler–Feint. To feint a strike at a high opening and change to another, usually lower, opening in the strike. Luring your opponent to commit to defending one opening and striking another.

Flech–German for the flat of the blade.

Fühlen–To feel. Feeling or gauging an opponent's pressure on your sword. That is, you feel his intent through the force and motion in his blade by feeling with your own. Always made in the bind. This is when you feel if he is hard or soft in the bind.

Gehiltz/Gehultz–The crossguard of the hilt.

Geschrenckt Ortt–A thrust in which the hands are held crossed over, the left hand under right.

Halb schwert–Halfsword. Techniques of gripping the middle of the blade itself with one hand (typically the left hand) while the other retains a hold on the hilt. This will allow for a wide range of offensive and defensive striking and deflecting actions as well as thrusts. It can also be used to thrust with great force and improved accuracy.

Hart und weich–Hard and soft. The idea that whenever contact is made with the swords (*im band,* in the bind) you gauge the pressure that the opponent places upon your blade (strong or weak). You then strive to oppose strength with weakness and weakness with strength in order to control and exploit your opponent's weaknesses in the bind.

Hende trucken–Pressing the hands. A term of Liechtenauer's meaning stop cuts to the opponent's hands or forearms. Done from below and then changed into a cut from above as you push him away. The term refers to the pressure of the hands exerted by means of the sword blade against the arms of the opponent.

Hert/Hart–Pressing with force in the bind.

Das Hängen/Hengen–The hanging/hanger. An important longsword movement referring to positions or actions where the point hangs downward from above. It can also be done from below by hanging the pommel to the ground; the point will then be pointing up at your opponent's face. From *hengen* you can do all the techniques described, but *hengen* as such is not a technique but a transitional movement of positioning the sword. The high and low hengen corresponds to the *ochs* and *pflug*, respectively.

Huten/Leger–The various guards or positions from which to fence well. In Ringeck's system of Liechtenauer they are the *vom tag*, *alber*, *pflug*, and *ochs*. In addition, there are the *nebenhut* and *schrankhut*. Later masters added to this list.

Indes–Simultaneously/at once. The concept of the moment when you can act and take initiative from your opponent. The idea of reacting without thought to an opponent's intent and changing thus from the *nach* to the *vor* and regaining the initiative.

Knopf–The pommel.

Krieg–War. The term means when the swords are in the bind and you work mainly with the point in order to hit your opponent or with the cut. Krieg is to be understood as the combined term for *hengen*, *winden*, *ringen*, and all other close-in techniques done from the bind.

Kron–Crown. The *kron* is a defensive action where you stifle a strike from above or turn a thrust to the side or up by lifting the sword high point up, or lifting it and tipping it to the outside, letting the opponent's blade slide off to the side.

Krumphaw/Krump–Crooked/twisted cut. This consists of a downward strike with the long edge executed with crossed or twisted wrists, or any strike with crossed hands. One of the *meisterhaw*. It counters the guard of the *ochs*.

Kunst des Fechtens–The German medieval (and Renaissance) art of fighting, consisting primarily of the arts of the *langen schwert* (longsword), the *messer* (a kind of falchion), and *ringen* (wrestling). Unarmored combat was known as *blossfechten*. Combat in plate armor was known as *harnischfechten*.

Kurtze schnyde–Short edge. The back or upper edge of the sword. It is the edge that lies between the thumb and index finger of your leading hand.

Lange schnyde–Long edge. The forward or true edge of the sword, on the side where your knuckles are.

Langer ort–Long point. A position assumed to create distance to your opponent when you are too close. Also to thrust with outstretched arms, not a good position to fence from since you are vulnerable to counters to your blade.

Meisterhaw–Master strikes. The very best techniques described by Liechtenauer, in essence the meisterhaw include *krumphaw*, *zornhaw*, *zwerchaw*, *schaytelhaw*, and *schielhaw*.

Mittelhaw–Middle strike. A left-to-right (or right-to-left) horizontal strike at middle height.

Mutieren–A *winden* variant. To direct the course of attack from the upper to the lower openings. *Winden* is a central concept in Liechtenauer's system, since *winden* is what allows you to change angle without any unnecessary movements, always presenting a threat with your point. *Mutieren* is special since you go over the opponent's blade to thrust instead of changing angle to his sword but not crossing over it.

Nach–After. Opposite of *vor* (before). If the opponent attacks first, you are left in the *nach*; that is, you must react to his action instead of forcing him to adapt to your actions. Ringeck states that you must push for the initiative at all times and regain *vor.*

Nachrayßen–Traveling after. One of the three ways of overcoming an opponent's attack by attacking him in the middle of his own attack or just after it is completed, usually by going after where the sword is (which makes it hard to deflect) or by going high if he is low. The concept encourages you to retake the initiative. But remember also that waiting for him to move may be advantageous to you; then, even if he moves first, you have the *vor.* The initiative is yours.

Nebenhut–Low or under guard. Ringeck states that it is best done on the left, but it works also on the right. In other works known as tail guard.

Oberhaw–Upper strike. Any strikes made from above your waist, either diagonally (*zornhaw*), vertically (*schaytelhaw*), or a *zwerchaw,* for example.

Oberschnitt–Downward slices made from above. See *Schnitt.*

Ochs–Ox. One of the two upper *leger* or *huten,* it can be done on the left or right side. In this position, you hold the weapon next to your head, with the point at your opponent's face. This guard can be done in variations with the sword straight and horizontal or sloping down like the horns of an ox.

Ort–Point. Also spelled *ortt.*

Pflug–Plough. One of the two lower guards, or *huten.* A stance of a bent position, it can be done low or almost upright, with the arms and the hilt close to the body next to the upper thigh and the blade at a 45-degree angle pointed at the opponent's face. It is well suited to making thrusts and defending against middle cuts. This is the fundamental "middle" stance and may appear to be more "left" or "right" depending upon which leg leads. However, the arms are only extended when an action or counteraction is executed. The posture is similar to the position taken when walking behind a yoke or a plow, hence giving the stance its name.

Redel/Rad–Wheel. A movement where you move the sword in a circle in front of your body from left to right with the right arm outstretched as you close in on your opponent.

Ringen am Schwert–Wrestling at the sword. Close-in techniques based on a few key actions: grabbing your opponent's hilt or arm, striking him with your pommel or guard, or trapping his forearms with your second arm. Several kinds of cuts could also be done, as well as kicking, striking, and using the elbow, throws, and grappling and disarming techniques known as *schwertnemen,* "sword taking."

Schaytelhaw/schayttler–Scalp strike/parting strike. A short and very quick, high horizontal strike aimed at you opponent's forehead delivered with your hands and elbows held high. It can also be a vertical downward strike aimed at the crown of the head. A possible interpretation of why it is called the parting strike is that you deliver it while stepping back to void the opponent's attack. It is always delivered with the long edge. One of the *meisterhaw.*

Schnappen–To snap. A sudden movement of the weapon, you use the opponent's pressure on your blade to let your sword snap around and hit him. It can only be done when he binds your sword hard.

Schnitt–Cut. These can be done as draw cuts or as pushing cuts in a forward direction. One of the three wonders, there are the upper and lower cuts. Cuts are usually aimed at the opponent's arms or hands, but may at times be aimed at his face or neck. A cut is almost always done from the bind after taking control of your opponent's sword by using the principle of weak and strong (using *indes* to find the opportunity) or by using the *nachrayßen* and reaching him with a cut to preempt his attack.

Schrankhut–Fence guard. Leading with either leg, the sword is held in front of the body on the same side, pointing down. The arms are uncrossed with the left foot leading and crossed with the right foot leading. Strikes are made while stepping forward.

Schwech–Weak. The German masters divided the longsword into two parts, a weaker section of the blade from the middle of the sword to the point was known as schwech. This was the part used for delivering damage through strikes, thrusts, and cuts. But it was a part of the sword where you preferred not to bind due to the weakness stemming from the distance from the hands and body.

Schwertnemen–Sword taking. Disarming techniques delivered in close quarters.

Shielhaw/Schiller–Squinting strike. A downward strike with the short edge at your opponent's shoulder or neck. The name is supposed to derive from the way that you look at your opponent with one eye only, squinting from the side; this is due to the manner in which your head and torso turn in delivering the blow. One of the *meisterhaw.*

Sprechfenster–Talking/speaking window. A position that Ringeck states is very versatile for discerning the opponent's actions. From the *sprechfenster* you can attack in numerous ways especially by using the *winden* and *durchwechseln.* It is usually done from the bind when the opponent's intent is spoken to you through his blade.

Stark/Starck–The stronger part of the blade closer to the hilt, this was used for parrying and binding. When it says that you work in the bind (*im band*), then the swords are crossed at the strong. You would strive to place the strong of your sword against the weak of your opponent's in order to dominate his blade and then use *winden* to place a thrust, for example.

Stücke–Piece. Techniques or combinations designed to get past an opponent's defenses.

Überlauffen–Overrunning. A timed counterattack where you outreach your opponent as he attacks. You would then move into or out of his attack and strike his closer target exposed by his movement. An attack to the lower openings is not parried, but voided by stepping simultaneously with a counterattack high, which will hit sooner than the lower attack.

Umschlagen–To pull away after a blow and strike at the opposite side.

Untere ansetzen–Thrusts or techniques delivered under or below the opponent's guard

Unterhaw–Lower strikes. Upward or rising strikes made from below the waist. They can be made diagonally as a *zornhaw* or vertically as a *schaytelhaw.* Usually made with the long edge.

Unter schnitt–To cut upward from below.

Verkehrer–A technique in which you bind the adversary's sword with an *oberhaw* and turn your sword so that your thumb is down and thrust to his face from above. This will force him to deflect the thrust to avoid being struck. You then step in and grab his right elbow with your left hand and place your left leg in front of his right leg and to toss him over.

Versetzen/Versatzung–Displacement. A defensive action to put off an attack by deflecting the incoming blow, or to counterstrike as opposed to an opposition block. All *versatzungen* are strikes, according to Ringeck.

Vom Dach/vom Tag–From the roof/from above. The basic position is with the sword by the right shoulder and right foot leading. It can also be done with the sword high over your head.

Vom Schwert–From the sword. An action done from the bind of the swords, when you do a technique in the *krieg*.

Vor–Before. The principle of being the first to act, thus forcing the opponent to react and robbing him of the initiative in the exchange of blows. *Vor* does not always mean being the first to act; sometimes it means not being the last to finish. It is the opposite of *nach*, or acting after the opponent does. A good swordsman can turn *nach* into *vor*.

Winden–Winding/turning. *Winden* is the action of turning the blade from side to side and from high to low and low to high while changing the angle of attack in the same movement. A central piece in Liechtenauer's system it can be done both *im band* as well as out of the bind. *Mutieren* is a special case of *winden*, since you cross over your opponent's blade, something you never do in *winden*.

Zornhaw–Rage strike/strike of wrath. A diagonal strike done from the shoulder or pulled back behind the back and executed with a step forward or back. It can be aimed from the head to the ankle. Also any diagonal strike, and of the *meisterhaw*.

Zucken–Jerk/twitch/draw. Striking in a side-to-side motion, a wind-milling/moulinet-type action.

Zufechten–The approach. The initial phase of combat. One of the two phases of combat where the combatants are closing distance to engage and an attack is made or their weapons make contact

Zwerchaw–Transverse strike/cross strike. A horizontal or slightly slanting strike made right-to-left. The *zwerchaw* is one of the *meisterhaw* and is used to deflect incoming strikes from above.

Bibliography

We have included not only the works that we have used in the making of this book, but also others that we feel is worth reading and that touch upon one aspect or another of swordsmanship and European fighting arts in general.

Amberger, Christoph J. *The Secret History of the Sword.* Unique Publications, 1999.

Angelo, Sydney. *The Martial Arts of Renaissance Europe.* 2000.

Fiorato, Veronica, Anthea Boylston, Christopher Knusel (editors). *Blood Red Roses: The Archaeology of a Mass Grave from the Battle of Towton AD 1461.* Oxbow Books, 2000.

Brown, Terry. *English Martial Arts.* Norfolk, England: Anglo-Saxon Books, 1997.

Clements, John. *Renaissance Swordsmanship.* Boulder, Colo.: Paladin Press 1997.

Clements, John: *Medieval Swordsmanship.* Boulder, Colo.: Paladin Press 1998.

Ducklin, Keith, John Waller, Adam Des Forges. *Sword Fighting: A Manual for Actors and Directors.* Applause Theatre Book Publishers 2001.

Hils, Hans-Peter. *Meister Johann Liechtenauer's Kunst des langen Schwertes.* P. Lang 1985.

Hope, Sir William, Donald McBane (ed.), Mark Rector (ed.), Paul Wagner. *Highland Swordsmanship: Techniques of the Scottish Sword Masters.* Chivalry Bookshelf 2001.

Jackson, James Louis (compiler). Giacomo Di Ragione Di Adoprar Sicuramente L'Arme . . . English, Grassi, Vincentio Practice, 1972. Saviolo, George (*Paradoxes of Defense*) Silver. Three Elizabethan Fencing Manuals. Scholars Facsimilies and Reprint 1972.

Kaufman, Stephen, et al. *The Martial Artist's Book of Five Rings: The Definitive Interpretation of Miyamoto Musashi's Classic Book of Strategy.* Charles E. Tuttle Co., 1994.

Keen, Maurice Hugh. *Medieval Warfare: A History.* Getty Center for Education in the Arts; illustrated edition, 2000.

Kurz, Thomsa. *Stretching Scientifically: A Guide to Flexibility Training.* Stadion Publishing Co., 1998.

Oakeshott, R. Ewart. *The Archaeology of Weapons: Arms and Armour from Prehistory to the Age of Chivalry.* Dover Publications, 1996.

Oakeshott, Ewart: *The Sword in the Age of Chivalry.* Boydell Press, 1998.

Prestwich, Michael. *Armies and Warfare in the Middle Ages: The English Experience.* Yale University Press, 1996.

Talhoffer, Hans, Mark Rector, John Clements. *Medieval Combat: A Fifteenth Century Illustrated Manual of Swordfighting and Close-Quarter Combat.* Greenhill Books/Lionel Leventhal, 2000.

SIDDORN, J. KIM. *Viking Weapons and Warfare.* Tempus Publications, 2001.

STUDER, CHARLES, ed. *Das Solothurner Fechtbuch.* Zentralbibliothek Solothurn. No date.

SUTOR, JACOB. *Kunstliches Fechtbuch* (No date. ISBN 3-928906-05-4. A very good reprint by a German publisher.)

THIMM, CARL A. *A Complete Bibliography of Fencing and Duelling: As Practiced by All European Nations from the Middle Ages to the Present Day.* Pelican Publications, 1999.

THORDEMAN, BENGT. *Armour from the Battle of Wisby.* Chivalry Bookshelf, 2001.

TOBLER, CHRISTIAN. *Secrets of Medieval German Swordsmanship: Sigmund Ringeck's Commentaries on Liechtenauer.* Chivalry Bookshelf, 2002.

UNDERWOOD, RICHARD: *Anglo-Saxon Weapons and Warfare.* Tempus Publications, 2001.

FLACH, ANDREW, Peter Field Peck. *The Official United States Navy SEAL Workout.* Hatherleigh Press, 1998.

WEALE, ADRIAN. *Fighting Fit: The Complete SAS Fitness Training Guide*, 1997.

WEINMANN (publisher): *Chronik alter kampkfkünste.* Berlin, 1997 (reprint of 15th–17th century wrestling manual, ISBN 3-87892-031-8).

WIERSCHIN, MARTIN: *Meister Johann Lichtenauers Kunst des Fechtens.* Munich, 1965

MANUSCRIPTS

DÖBRINGER, HANKO. Codex MS 3227a. Germanisches Nationalmuseum, Nürnberg.

RINGECK, SIGMUND. *Fechtbuch.* MS Dresd.C 487. State Library of Saxony, Dresden.

Online Resources

FOR THE BENEFIT of those who are not yet well acquainted with the world of European martial arts, we have included a small list of useful resources. To find more links look at their rather extensive link lists.

ORGANIZATIONS

ARMA (The Association for Renaissance Martial Arts)–Teaches almost every aspect of European martial arts. Its Web site is a gold mine of information and chapters exist worldwide. Good articles and research. <www.thehaca.com>

DIE FREIFECHTER–German teaching organization of very high standard. Good resource for German manuals, both transcriptions and comments. <www.freifechter.org>

EUROPEAN HISTORICAL COMBAT GUILD–English teaching organization of good standard. Chapters in other European countries. Does some research and publishes on its Web site. <www.ehcg.org>

SWORD FORUM INTERNATIONAL–Very good site that has a very lively chatboard, and a good place to find practice groups or people who can answer your questions on the topic of swordsmanship. <www.swordforum.com>

SWORDS

ALBION ARMORERS–Excellent maker of good swords in almost every style. From Roman to Renaissance, well crafted and made to be used for test cutting. <www.albionarmorers.com>

PURPLEHEART ARMOURY WOODEN SWORD WASTERS AND WEAPONRY–Excellent wasters of every kind, the best we have tried. Warmly recommended. <www.woodenswords.com>

JELLING DRAGON: ROMAN, VIKING AND MEDIEVAL REPLICAS–Sells Paul Binns swords, which are very durable for practice, but slightly too heavy, but that is good for your deltoids! <www.jelldragon.com>

LUTEL–Czech swordsmith who makes very good practice blades at decent prices. <www.lutel.cz>

ARMOR

VIA ARMORARI–Excellent armor, made to be used in steel–and they also temper the pieces! Good prices, and their craftsmanship really works when fighting with steel replicas. <www.viaarmorari.com>